Zero Pollution for Industry

ZERO POLLUTION FOR INDUSTRY

WASTE MINIMIZATION THROUGH INDUSTRIAL COMPLEXES

NELSON LEONARD NEMEROW

A WILEY-INTERSCIENCE PUBLICATION

JOHN WILEY & SONS, INC.

NEW YORK / CHICHESTER / TORONTO / BRISBANE / SINGAPORE

Library of Congress Cataloging in Publication Data:
Nemerow, Nelson Leonard.
     Zero pollution for industry : waste minimization through
industrial complexes / by Nelson Leonard Nemerow.
        p.   cm.
    Includes index.
    ISBN 0-471-12164-9
     1. Factory and trade waste—Management. 2. Waste minimization.
3. Recycling (Waste, etc.) 4. Industrial sites. I. Title.
    TD897.5.N46 1995
  363.72'8—dc20   95-10072

Almost by the nature of their concerns, economists focus on goods and services as the primary focus of attention, with environmentalism as a secondary issue. Environmentalists see the world exactly in reverse. A clean environment is primary: more goods and services are secondary.

—LESTER THUROW
"Head to Head, 1992"

The economy is a wholly owned subsidiary of the environment. All economic activity is dependent upon that environment with its underlying resource base. When the environment is finally forced to file under Chapter 11 because its resource base has been polluted, degraded, dissipated, and irretrievably compromised, then the economy goes down with it.

—GAYLORD NELSON
"Earth Day, 1970"

CONTENTS

After more than 50 years of active participation in all aspects of the industrial waste treatment field, I have concluded that governmental regulation alone will not alleviate the environmental damage caused by industrial wastes. Some consideration must be given to the economic consequences of these waste discharges. For at least half of my years of involvement, I have advocated putting a monetary value on industry's use of any natural resource for its waste consumption. Up to now society has found such a monetary system too cumbersome, impractical, or difficult to implement. Some breakthroughs are beginning to emerge, however. For example, the federal government is cooperating in a marketing system for limiting the amount of sulfur dioxide discharged into the air resource by power plants. For years, municipalities have been charging industry for sending certain of their contaminants to their treatment plants through community-owned sewers. However, neither these nor other attempts to charge industry have included costs of the potential environmental damage resulting from these wastes. Most roadblocks to such considerations have been due to our inability to measure accurately the costs of such damage. But we are gaining knowledge and understanding of the economics of adverse environmental impacts of wastes. In the not too distant future, I predict, we will avoid industrial pollution mainly through putting a price on the usage of our diminishing natural resources of air, land, and water.

In the meantime, industry possesses a valid alternative to avoid these costs. In fact, if such systems as I am proposing and describing in this book are implemented, societal as well as industrial costs for pollution abatement may be eliminated altogether. Briefly, the system entails (1) waste minimization, (2) waste reuse, and (3) industrial complexing to achieve zero pollution.

The first chapters of this book present the concepts and some details about reducing industrial wastes to a minimum and reusing industrial wastes within the plant and by others to a maximum. Both approaches will reduce contaminants considerably but will not eliminate all pollutants completely. Chapter 5 describes in detail the system of environmentally balanced industrial complexes. Such complexes eliminate all industrial contaminants from our natural environments. To use this system, one must be innovative and courageous. The complexes are designed to include compatible industrial plants adjacent to one another so that all wastes produced by one are used completely by another plant to make its products. In this way, no wastes will enter the environment and industrial production costs will be reduced.

Many examples of this concept are given in Chapter 5; some with more data than others. Some complexes have been researched, while others are mainly conceptual. The intent is to offer the reader a beginning plan for avoiding the costs of industrial pollution. Much more investigation, research, and additional pilot studies must be launched and continued in the effort to attain zero pollution.

I have made a special effort to make the Index of this book useful to readers of all disciplines. Therefore, I recommend using the Index as a first process for locating subjects of interest to you.

I want to thank Catalina Hope, Cathy Richards, and Hal Richards for their valuable word-processing time and experiences in preparing the first and second versions of the original manuscript.

Appreciation is due to Daniel R. Sayre, Senior Editor, and Tracey Thornblade, Assistant Editor, at John Wiley for their confidence in the manuscript and perseverance and patience with me in the publication of it.

Finally, I want to acknowledge the brainstorming session I had with Alex Anderson of the United Nations Industrial Development Organization in Vienna, Austria, during the early 1970s, at which time the Industrial Complex idea of Chapter 5 was born.

Zero Pollution for Industry

CHAPTER 1

RATIONALE FOR ATTAINMENT OF ZERO POLLUTION

IMPORTANCE

Our fragile and limited environment is rapidly approaching the breaking point. No longer can industry discharge any amount of contaminants into the surrounding air, water, and land without some adverse effects. This was not always the case. In years gone by, there were some contaminant-carrying resources available for the wastes of industry discharged into our environment. Most waters contained some dissolved oxygen; most air freely diluted noxious gases and dispersed them over uninhabited areas; most lands were far from people and drained into distant waters with adequate dilution. Today we are faced with deficiencies in most of these environmental resources, as well as increased concentrations in habitation and greater productive land use everywhere. Wherever industry locates today, a growing population is likely, accompanied by multiple usages of the environmental resources of land, water, and air. For some years in the past, we struggled with the proper allocation of dwindling resources. Now we have arrived at the point of needing to prohibit any degradation of these resources by industry. Our world population and its demand for products has increased so fast and to such an extent that there is little or no room for tolerance of waste discharges. The very people demanding more industrial products and services also need and require these land, air, and water resources for their very survival. For without clean air and water and uncontaminated land, there will be no quality of life for habitation as we have known it. People will be forced to wear gas masks with protective filters to purify the air, to carry or otherwise depend on ubiquitous water purifiers to provide safe water for drinking, and to use contaminant-testing equipment whenever venturing into unfamiliar land areas. Difficult as this may be to imagine, it will happen if we humans are willing to avoid our responsibility to curtail all industrial pollution.

Environmentalists are no longer considered scaremongers and extremists. The true environmentalist of today is sincerely interested in preserving our standard of living and way of life. To do this, we must aim for zero pollution as a result of the manufacturing of any industrial product. I once advocated charging industry a price for using up our environmental resources, increasing the unit price as the resource diminished. This idea was proposed as a method of encouraging industry to prevent any excessive discharge of contaminants. It is still not too late to use this system, but the unit prices for resources have risen to such high levels that even this is becoming prohibitive. When it is not possible to put prices on environmental resources, we must use stronger, more prohibitive measures to avoid resource depletion and the resulting catastrophes.

THREE GENERAL METHODS TO ATTAIN ZERO POLLUTION

Industry must eliminate all pollutants from its effluent in order to attain zero pollution. Essentially, there are three methods to eliminate all pollutants. They are presented in order of probable acceptance by the industry. Acceptance is related to the ease of implementation by any industrial plant. The first method is *recovery and reuse within the same plant* to minimize waste to the greatest degree possible. Under optimum conditions, this procedure could lead to zero pollution by an industrial plant. More likely, however, a plant will attain a 25% to 75% reduction in contaminants by recovery and reuse. A typical example of this is paper mills' use of save alls to recover machine *"white water" waste material.* The second method is the *recovery and sale of wastes to other manufacturers* outside the industrial plant. An example is the recovery of fleshings from tanneries for resale and reuse by rendering plants to manufacture grease, soaps, and animal glue. Another possibility offered by this method is for a plant to sell its waste to a regional exchange, which then contracts with a third party for sale and reuse. The exchange acts as a middleman contractor.

The third method is an obvious and logical result of integrating both the advantages and the disadvantages of the first two procedures. This method reaches zero pollution by *bringing the waste producer and user together in one industrial complex.* The user avoids the transportation costs usually associated with bringing in raw materials for processing. The producer saves the money usually associated with treating its wastes to protect the surrounding environment. Because the user must have an operation that is compatible with the producer's, the proper selection of compatible industrial plants is necessary to ensure a workable and efficient industrial complex. A typical example might include both a tannery and a rendering plant in a complex so that the tannery would not have to dispose of its fleshings and the rendering plant would have a ready, on-site source of fleshings as raw materials.

However, before we begin any program involving the reduction of wastes to a minimum, we must understand the overall economics of pollution. For, if it were not for the cost involved, there would be no resistance by industry to provide pollution control. Therefore, it is only logical to conclude that money lies at the very root of the problem of eliminating industrial pollution. In fact, a plan to encourage cost minimization of pollution abatement would enhance its acceptance by industry. In the next chapter we consider some ways of using cost savings to encourage industrial pollution abatement.

CHAPTER 2

ECONOMICS OF ZERO POLLUTION

NECESSITY FOR WASTE TREATMENT

At one time it was necessary for a manufacturing plant to treat its waste only if and when it violated criteria and standards established by a governmental regulatory agency. Even then, industry resisted compliance in order to avoid costs that would jeopardize profits. Today, in most instances, industry consents to waste treatment in order to preserve environmental quality for all users surrounding its plants. Slowly, but surely, industry worldwide is accepting waste treatment as an integral part of production costs. The added costs must then be passed on to consumers or deducted from the profits of the firm. In a competing market, industry must now make decisions on methods of minimizing waste treatment costs (environmental costs).

COMPONENTS OF PRODUCTION COSTS

Production costs in all industries comprise three components:

1. Operation
2. Maintenance
3. Materials

Although these three costs vary from one location to another and from one time to another, they can generally be predicted at the onset of production with some degree of certainty. Because they cannot be predicted precisely, they are often referred to as variable costs.

On the other hand, industry must also include capital costs in its decision making. Because these costs can be predicted with certainty at the beginning of a project, they are referred to as fixed costs, which include:

1. Depreciation
2. Taxes
3. Net interest charges

Fixed costs vary according to the type of industry, location, time, and market for the product. It is rather difficult to make a generalization for fixed costs.

Therefore, for the purpose of this discussion, we will assume that the decision has been made to manufacture a given product and to include ample consideration of fixed costs in its selling price. Our goal is to clarify the relationship between production costs and waste treatment costs.

GENERAL WASTE TREATMENT COSTS

The direct cost of waste treatment is more than just the expense of operation, maintenance, and materials involved in using capital equipment to render industrial waste harmless to the environment. This direct cost represents only a portion of the total cost. It is a readily measurable quantity' whereas the other, indirect portion may not be as easily identified and quantified without considerable debate. Regardless of how waste treatment costs are ascertained, reported, or used, it is absolutely vital that both the measurable (direct) and the less measurable (indirect) costs be identified, included, and used. Indirect costs are those related to the adverse effects of the wastes on the environment. Indirect costs are external costs to an industry, in that they must be borne by society often at some distance from the manufacturing plant. Some can be quantified, but most are costs of damages, which are open to subjective evaluation. We can conclude generally, however, that the less the waste treatment provided by an industry, the greater will be the costs of damage to the environment. Damage costs generated by waste will also depend on the present and anticipated future use of the air, water, and land affected by the discharge. If no waste treatment is provided by the industry, environmental damage costs will be maximum. Such is the assumption made by T. N. Veziroglu[1] (1992) in computing the damage costs for fossil fuel power plants to be 32.18 mils/kwh.

COST OF INDUSTRIAL WASTE TREATMENT

The capital and operating costs of industrial waste water treatment must be computed, clearly expressed, and presented to management in a readily understandable form. Once the mysterious so-called external costs are identified, acceptance by an industry becomes much easier. I recommend that all industrial waste treatment costs be calculated on an annual basis to include both capital and operating expenses. This makes them comparable to other industrial manufacturing costs such as labor, transportation, utilities, rent, marketing, and even administrative charges. Open comparison of waste treatment with these other production costs can result in enlightened and rational decisions.

In the past, industry used two methods to express the cost of waste treatment; one for public relations, to discourage implementation; the other for internal propaganda, also to discourage implementation. In the first instance, the cost is presented simply as a lump-sum, present-day charge. When submitted to society as a single, unrelated, and presumed unnecessary cost, it sounds absolutely unreasonable. For example, $5 million for power plant cooling towers would obviously appear excessive. In the second approach, the cost of waste treatment

---

[1] T. N. Veziroglur, H. J. Plass, Jr., and F. Barbir, *Electrochemistry in Transition* (New York: Plenum Press, 1992), "Hydrogen Energy System: Comparison with Synthetic Fossil Fuels," 325–328.

is expressed as a cost per share of stock, to be subtracted from low or already diminished corporate earnings. When industrial managers, corporate board members, and stockholders consider these expenses and their consequences, they naturally react negatively to implementing waste treatment.

It is much more realistic and fair to express the cost of wastewater treatment as a charge per unit of production cost. This charge can be considered as "value added." Therefore, an appropriate expression would be cents of waste treatment per cents of value added for an industrial product, or percentage of value added. Such information would provide industry with a direct comparison of its cost of waste treatment with other direct costs. It may also provide clues as to how to pay for these costs once a decision is made to proceed. In many instances, the cost per unit of production is so small that it can readily be absorbed by an industry without raising the market price of its product. In addition, this approach may force the industry to seek alternative methods of treatment, through ingenious innovation, to lower production costs in order to maintain a constant profit level. Another method I suggested is the sales index, that is, waste treatment cost as a percentage of an industry's sales volume.[2] Sales records, generally public information, are usually readily available, whereas production costs may be difficult to obtain and assess.

Whether costs of waste treatment are related to production costs or sales, the resulting percentages are best compared on an industry basis. For example, in the pulp and paper industry I reported that the average cost of waste treatment (85% BOD removal) was found to be 4.34¢ per

---

[2] N. L. Nemerow, "Benefit-Related Expenditures for Industrial Waste Treatment," Bulletin, on Cost of Water Pollution Control National Symposium, Raleigh, North Carolina, 1972.

sales dollar (1972). Paper mills incurring waste treatment costs greater than this may be extremely hard pressed to comply with EPA guidelines. In another unrelated type of industry, however, a sales index of up to 4.34¢ may be too high for any plant within that industry to afford. In the nitrogen fertilizer industry, moreover, the ratio is 0.5% for ammonia, ammonium nitrate, and ammonium sulfate. The figure for urea is about 2.0%, and for phosphate fertilizer plants ratios vary between 0.5% and 0.3%.[3]

For a number of years, I have been collecting reliable economic data from certain industrial plants. In these cases I attempted to relate waste treatment cost to value-added or production cost. Some of the percentages are included in Table 2.1

I also expressed the cost of waste treatment as dollars per million gallons of wastewater treated.[4] This is a conventional method of expressing treatment costs for domestic sewage, but because of the variations to be considered (as mentioned early in this chapter), it is not highly recommended for industry.

We hear a great deal these days about not being able to afford to pay for pollution control. We hear that we cannot have both jobs and capital spending for growth and, at the same time, clean air and water. For the equivalent of secondary-type treatment, these statements are simply untrue for industry. As seen from the relatively low cost of industrial waste treatment shown in Table 2.1, these arguments are not valid.

All but the unessential, borderline, or unprofitable industrial plants can afford waste treatment costs, and should not attempt to bypass them with subterfuges of one

---

[3] J. Carmichael, *Environmental Problems in the Fertilizer Industry* (Vienna, Austria: UNIDO, 1978), 12.

[4] N. L. Nemerow, *Industrial Water Pollution: Origins, Characteristics, and Treatment* (Malabar, Fla.: Robert E. Krieger, 1987), 78.

Table 2.1 Industrial Waste Treatment Costs

Industry Type	Type Production	Waste Treatment Cost per Dollar Value Added (%)	Data Originator
1. Cannery	Beans	1.11	Nemerow, 1972
2. Cannery	Tomatoes, peaches	1.08	Nemerow, 1972
3. Poultry Processing	Chickens	1.096	Nemerow, 1970
4. Tannery	Upper leather	0.42	Nemerow, 1972
5. Iron and Steel	Finished steel	1.8	Nemerow, 1976
6. Petrochemical	Ethane-propane cracking	0.5	Carmichael, 1977
	Naphtha cracking	0.9	Carmichael, 1977
	Gas-oil cracking	1.4	Carmichael, 1977
	Ammonia via natural gas	1.1	Carmichael, 1977
	via coal gas	7.2	Carmichael, 1977
	via heavy oil gasification	3.1	Carmichael, 1977
7. Fertilizer	Nitrogen Fertilizer	Max. $0.78/ton product	Carmichael, 1978
	NPK plant	$2.15/ton product	Carmichael, 1978

Source: N. L. Nemerow, "Costs and Innovative Solutions for Industrial Waste Treatment," in *Treatment and Disposal of Liquid and Solid Wastes,* ed. Kriton Curi, (Oxford and New York: 1980) Pergamon Press, 469–473.

type or another. When an industry is forced to remove more contaminants than can be accomplished with secondary treatment, other solutions may be preferable to the more costly tertiary treatment.

SOLUTIONS TO INDUSTRIAL POLLUTION CONTROL

Treatment of wastes can eliminate, partially or completely, all environmental damage costs. A decision must be made as to what degree of treatment is required to arrive at the optimum outcome and damage cost control. An excess of treatment is not economically sound if its cost exceeds residual damage costs; on the other hand, too little or no treatment can result in excessive damage costs. The general relationship between treatment and damage costs is shown in Figure 2.1.

In this diagram, we have two separate industrial pollution situations, A and B. Separate treatment and damage costs curves are shown for each industrial problem. As environmental contamination and its associated cost approaches zero, treatment costs are a maximum and damage costs a minimum. The intersection of the two curves for each case represents the treatment required and associated costs, and passes on all environmental

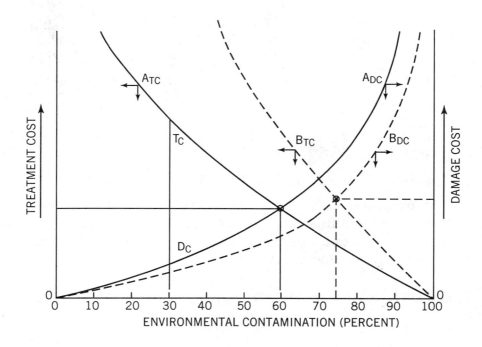

Figure 2.1. Environmental Protection Versus Treatment and Damage Cost

damage costs to society. Because of population growth and increasing land use, environmental damage costs are less acceptable to society. Treatment or alternate solutions must now be investigated.

There are four major overall methods that industries and municipalities can use to eliminate completely the discharge of contaminants into the environment—whether air or water, or both. There are drawbacks, as well as advantages, in the use of each of these methods. The discharger of such contaminants must weigh the pros and cons of each method as it relates to the producer's own particular situation. Economics will usually be the primary criterion in making the decision, but any cost must include the environmental monetary benefits of abating pollution.

The four generally accepted solutions to pollution abatement, along with pertinent generalizations of each, are included in Table 2.2.

One problem with recycling as a solution to environmental pollution is the cost of recovering and reuse of materials. Unless there is a demand for recycled materials in the quantities generated and for the types and qualities available, recycling costs may equal or exceed the original value of the materials. Here again, we must include the elimination of environmental damage as a negative cost of recycling.

For example, in south Florida it was reported that so much is being recycled that "it is starting to cost money." One city will start being billed $15 a ton for the removal of old newspapers, whereas three years ago the same city was paying $30 a ton. In fact, in Broward County, Florida, the average price paid for recycled newspapers dropped steadily from $35 a ton in 1988 to $3.50 in 1992. However, the alternative available to Broward County would have been burning the paper at a cost of $62.50 per ton.

Table 2.2 Solutions to Pollution

Solutions	How Accomplished	Advantages	Problems Associated
1. Prevention	An administrative decision is made to discontinue use of contaminant.	No further need for treatment, disposal, or reuse. Environment is protected from any deterioration.	A substitute chemical must be found, which may result in inferior or more costly production.
2. Treatment and Disposal	Contaminant is removed or destroyed from effluent stream by specific treatment *and* disposed of safely in adjacent environment or exported to an acceptable location.	No environmental degradation from contaminant.	Treatment may be costly and/or may incompletely remove the contaminant.
3. Recovery and Reuse Elsewhere	Specific contaminant recovered in unchanged condition and sold for reuse by another producer at another location.	Some monetary recovery from reuse may result. No contaminant released to surrounding environment.	Recovery may be expensive. A reuser may be difficult to locate.
4. Direct Reuse Within	Contaminant is collected and/or directed to another location within the complex for another use.	No treatment costs. No environmental damages. Reduced production costs effected by chemical savings.	A reuser or user must be identified within the complex. Contaminant may have to be modified before reuse.

At the same time, the report states, since 1988 the prices cities paid for recycled aluminum has dropped 50% from $.60 to $.30 per pound; for green glass, even more—from $50 a ton to $15; and for brown glass—from $55 to $25. Even the price for recycled clear glass (the most desirable color) has decreased from $55 to $50 a ton.[5]

---

[5] S. Borenstein and B. French, "Recycling Cost Go Up as Supply Outstrips Demand," *Sun Sentinel* (February 1992): 2 C.

All manufacturing products are influenced by certain market forces. These generally affect recycling costs as well.

Conventional Industrial Waste Treatment

The pollutants contained in municipal wastewaters are quite well known to most workers in the field of environmental engineering. They include solids, BOD, microorganisms, and the more recently recognized dissolved minerals. Liquid wastes of industries, however, may contain these and many more possible contaminants, such as coloring agents, toxic organic chemicals, heavy metals, oil, detergents, acids and alkalis, odors, heat, radioactivity, and many types of colloidal solids. In addition, industrial effluents are much more unpredictable as to both quantity and quality at any given time. The potential combination of various unpredictable contaminant at any time of the day, week, or season of the year makes proper and effective treatment very difficult.

Industry has at least 12 alternatives available to solve its wastewater treatment problem. These are illustrated in Figure 2.2.

The major paths open to an industry include treatment by the municipality of all its wastes, either before or after treatment by the industry, or treatment to some degree solely by the industry. The challenge is once again to determine which path industry should to follow. The problem can be solved by investigating the parameters of the situation. After the facts are revealed, they must be evaluated and integrated into appropriate models to yield the optimum solution. Although technical parameters such as quantity and quality of wastewater and type of municipal treatment available are the primary facts, economical, political, social, and psychological factors are also extremely important.

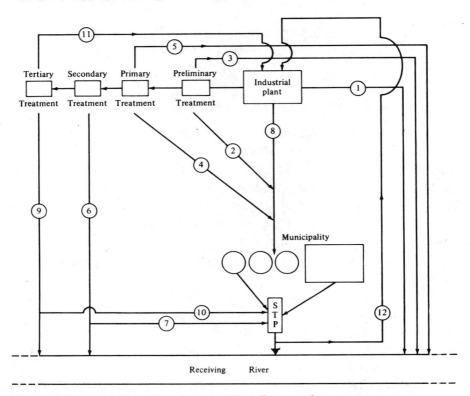

Figure 2.2. Twelve Alternatives of Industrial Waste Treatment Systems
Source: N. L. Nemerow and A. Dasgupta, *Industrial and Hazardous Waste Treatment* (New York: Van Nostrand Reinhold, 1991), 238.

Traditionally, two systems have been used for controlling industrial contaminants to an amount that can be discharged safely into our watercourses. One is manifested by establishing and maintaining water quality standards for the receiving streams. This prevents industry from putting more contaminants in to the receiving waters than they can tolerate. The standards selected will depend primarily on the stream use. The total quantity of contaminants allowed will also depend on the quantity and quality of stream flow available. Because of the variability of both stream flow and the industrial contamnants and the increasing number of contaminant contributors, this control method is difficult to apply. The second system uses a standard for industrial effluent concentration or quantity to limit excessive contaminants. Basically, each type of industry is expected to limit its discharge of common contaminants to a reasonable level. This method is much easier to apply and more equitable, but it also bears less correlation to conditions in the receiving water.

Because the stream standards proved so difficult to administer and the effluent standards so easy to monitor and control, greater consideration with some modification, was given to the latter. For these reasons, we are using a system of effluent guidelines for each of about 30 prime wet industries. Each is allowed to discharge only a reasonable percentage—possibly 15% to 50%—of its major and most common raw wastewater contaminants. It is too early to say whether this system will prevent excessive amounts of them from entering our streams. However, the individual states still retain the right to demand additional treatment by industry where the guidelines alone do not adequately protect the quality of specific receiving waters.

Some Modern and Innovative Solutions in Industrial Waste Treatment

Four modern and innovative solutions to the problem of industrial waste treatment are (or will become) available. These can be summarized briefly as (1) recovery and or reuse, (2) the tying of effluent levels to stream resources, (3) the marketing of stream resources, and (4) the construction of environmentally optimized industrial complexes. This chapter describes the meanings, problems of implementation, and likely results of these solutions.

Reuse—directly within a plant or indirectly by other industrial plants or municipalities—offers a ray of hope solving some of industry's problems. There are several obstacles to this ideal solution. First, a strong program in education must precede a successful reuse program. We should acquaint plant personnel with the potential value contained in wastewater. Second, a rather detailed qualitative analysis of wastes must be made available over a relatively long period of time. This information can fortify plant managers' convictions to reuse these wastes. Third, and probably most important, there must be an acceptable user nearby for the wastewater available. The recipient should have a water-use need closely correlated to the quantity and quality of the wastewater available from the donor industrial plant.

Industrial wastewaters vary in quantity and quality from time to time. As the type and amount of product changes, so does the industry's wastewater. The same is true of the contaminant-carrying capacity of the receiving streams. Most streams possess both high and low flow periods corresponding to the time of the year, rainfall, and topography. One possible way to avoid excessive waste treatment costs is to gear production costs at the industrial plant to the stream resources available at the time. When little stream flow and stream resource are available, industrial production can be reduced. Likewise, when there is ample stream flow and stream resource, plant production can be increased again. Long-term industrial production probably would be unchanged by this technique. Such controlled production is commonly used to accommodate raw material supply, market conditions, and labor available, and could likewise be applied to stream resources.

To implement this solution, a telemetering system, providing data from the stream to the plant could be installed to control production. However, both industry and water pollution control agencies would have to be convinced of the efficacy and utility of this method. Industry is often wary when another constraint is placed on its production. Regulatory agencies, on the other hand, doubt the honesty and integrity of industry to comply with agreed-upon production requirements.

One sure way to persuade industry to seek acceptable solutions to wastewater treatment problems is to penalize companies for violating the laws of nature. These laws are often different from the laws of humans. The law of nature dealing with wastewater states that you shall not discharge into a watercourse more contaminant than can be assimilated without an adverse environmental impact. A potential method for assuring society that an industry will not violate this law of nature is to charge it a fee for using the existing resource. A unit price for the stream resource roughly equivalent to the total benefit to society of that resource can be established. Then each consumer (industry, people, or agriculture) can purchase given amounts or units of this resource for that price. As stream resources diminish, higher unit prices could be established to discourage overconsumption leading to a state of pollution. In 1974 I proposed this system of protecting receiving water resources.[6] It has not been implemented so far, mainly because of the relatively inexpensive cost of waste treatment and the relatively large effort that must be expended to initiate the system.

The ultimate in solutions to industrial wastewater problems is one utilizing environmentally optimized industrial complexes. These complexes are designed to contain a number of compatible industrial plants, co-located and practicing endless reuse and recirculation of all residues: liquid, solid, and air.

The raw materials for each plant are intended to be the waste materials from other plants of the complex. In this

---

[6]N. L. Nemerow, *Scientific Stream Pollution Analysis* (New York:, McGraw-Hill 1974).

way, the environment suffers little or no adverse impact, and production costs are minimized. The key to the success of such complexes lies in our ability to select the appropriate combination of plants and production capacities for each, and to obtain industry's cooperation in the whole concept. I described this novel solution in the case of the pulp and paper industry in 1979.[7] Wet industries that may benefit from industrial complexing include, among others, the steel, fertilizer, tanning, textile, oil refining, petrochemical, and certain food processing industries.

Plants in these industries are typically so large that new plants are seldom constructed. This situation can delay implementation of complexing as a solution. Another obvious deterrent to this system is the problem of determining the optimum combination of plants for specific complexes. Further, the assistance of an industrial development agency would be helpful in obtaining agreement for locating plants of these types and sizes within complexes. Despite these hurdles, the benefits to be gained from such solutions far outweigh the efforts needed to achieve them. In fact, industrial complexing offers one of the most promising long-range solutions to today's environmental pollution problems, as well to as many industrial economic problems of the future.

Benefit-Related Expenditures for Industrial Waste Treatment

Industrial waste treatment is a necessity to preserve our air, land, and water resources. The economic stability of

---

[7]N. L. Nemerow, S. Farooq, and S. Sengupta, *Environmentally Optimized Industrial Complexes for the Pulp and Papermill Industries*. Paper presented at the Industrial Waste Treatment Conference for Pulp and Papermills in Calcutta, India, December 1977. Published in 1979 in *Environmental International* 3:1, 133.

our society is also vital to our well-being. Waste treatment may cost more than an industrial plant is willing, or even able, to spend. This is especially true in situations where environmental resources are limited and there is intense competition between users and consumers of these resources. Unfortunately, these situations are becoming more prevalent. What is an industry to do in such cases? Move? Cease production? Enter into a legal maneuver to delay or prevent excessive costs for waste treatment? None of these is really desirable for industry or for society. What should governmental regulatory agencies do in these critical situations? Force industry into one of the undesirable alternates or ignore the need to protect the resources and allow the plant to continue to pollute? Neither of these positions is satisfactory. How, then, do we solve the problem of apparently conflicting interests of two factions of our society? It is vital to the objectives of this book to resolve the conflict between the economics of industry and the general welfare.

EXPENDITURE JUSTIFICATION

Lest any party question whether wastewater treatment costs are justifiable, I present the following lists of primary, secondary, and intangible benefits.[8]

Primary Benefits

1. Savings in dollars to the industrial firm by reuse of treated effluent instead of fresh water
2. Savings in dollars resulting from compliance with regulatory agencies, for example, avoidance of legal and expert fees and time of management involved in court cases

---

[8] N. L. Nemerow, *Liquid Wastes of Industry—Theories, Practice, and Treatment* (Reading, Mass.: Addison Wesley, 1971), 51.

3. Savings in dollars from increased production efficiency, made possible by improved knowledge of waste-producing processes and practices

Secondary Benefits

1. Savings in dollars to downstream consumers resulting from improved water quality and, hence, lowered operating and damage costs
2. Increase in employment, higher local payroll, and greater economic purchasing power of the labor force used in construction and operation of waste treatment facilities
3. Increased economic growth of the area because of the commitment of industry to waste treatment and potential of expansion of the existing plant
4. Increased economic growth of the area, with more clean water available for additional industrial operations which, in turn, yield more employment and money for the area
5. Increased value of adjacent properties as a result of a cleaner, more desirable receiving stream
6. Increased population potential for the area, because cleaner water will be available at lower cost, the limiting factors of water cost and quantity having been pushed further into the future
7. Increased recreational uses, such as fishing, boating, swimming, as a result of increased purity of water; the availability of recreational opportunities previously eliminated

Intangible Benefits

1. Good public relations and improved industrial image upon installation of pollution abatement devices

2. Improved mental health of area citizens, confident of having adequate waste treatment and clean water
3. Improved conservation practices, which eventually yield payoffs in the form of more clean water for more people for more years
4. Renewal and preservation of scenic beauty and historical sites
5. Residential development potential for land areas nearby because of the presence of clean recreational waters
6. Elimination of relocation costs (for persons, groups, and establishments) because of impure waters
7. Removal of the potential physical health hazards of using polluted water for recreation
8. Industrial capital investment assuring permanence of the plant in the area, thus lending confidence to other firms and citizens depending on the output produced by the industry
9. Technological progress, resulting from conception, design, construction, and operation of industrial waste facilities

Obviously, these benefits must be quantified in some manner in order to arrive at a specific level of justifiable expenditure. At this point, I would like to express my opposition to the view expressed by some that all industrial waste treatment costs are justifiable to protect all stream resources. Advocates of this position make light of any attempt to quantify benefits because of their foregone conclusions. They further believe that wastewater resources engineers are poaching on other fields in applying economic measures to treatment decisions. What these overexuberant conservationists fail to consider is that our economic ability to ameliorate society's ills is limited. We simply cannot afford to do everything to improve the environment instantly. Therefore, someone has to establish priorities. We are obliged to provide government with formulas, or at least methods, for making more objective decisions in pollution abatement.

QUANTIFICATION OF BENEFITS

We can begin quantification by defining benefits as the amounts actual and potential water users are willing to pay or the value of avoiding payment of a given number of dollars at a given quality. The dollar benefit of a water resource at a given quality may be determined by listing all the uses that are affected by water quality, by valuing each use individually, and by summing the resultant values. The major uses affected by water quality are grouped in the following categories: (1) recreation uses, (2) withdrawal water uses, (3) wastewater disposal uses, (4) bordering land uses, and (5) in-stream water uses. The value of these uses may be estimated by taking surveys of the users to determine the extent of demand for each use and the amount each user is willing to pay for a unit of use, or the user benefit. Annual dollar benefits for a given use are the product of the total demand times the unit benefit. Total annual dollar benefit at a given water quality is the sum of these benefits for each use.

Total annual dollar benefit at a higher water quality may be estimated by determining the probable demand for beneficial water uses at the new quality. This demand may be estimated by surveying the present need for comparable uses at a nearby lake or stream with this new quality. It may also be estimated by questioning potential water users to determine the latent demand for water at this new quality for possible beneficial uses that are presently being foregone. Discussion of these beneficial uses follows.

Recreation Use Benefits. Water-oriented recreation uses include sightseeing, walking and hiking, swimming, fishing, picnicking, boating, hunting, camping, waterskiing, canoeing, sailing, and skin and scuba diving. These recreation uses may be valued by including all the expenditures of the average recreationist as a measure of his or her willingness to pay. These include the costs of equipment, food, travel, and recreation area user fees.

Withdrawal Water Use Benefits. Withdrawal water uses include municipal water supply, industrial water supply, and agricultural and farmstead water supply. The water quality benefits reflected in a municipal water supply may be estimated to be at least equal to the cost of water treatment by chemical coagulation, sedimentation, and rapid sand filtration. Water quality benefits for an industrial water supply may be estimated from water treatment costs, not to exceed those for municipal treatment. Industrial costs to produce ultrapure process water are not assigned as water quality benefits, because these costs are related more to overhead costs of a particular manufacturing process in contrast to the cost of a normally supplied public utility. Agricultural and farmstead water use benefits may be estimated as negative values if damages have occurred to irrigation, poultry and livestock watering, farmstead family, or dairy uses.

Wastewater Disposal Benefits. Wastewater disposal benefits may be estimated as the total annual costs for waste treatment required to meet existing standards for minimum stream or effluent standards. The difference in annual costs between the existing level of treatment and the level required to meet the minimum standards may be considered a present benefit to the waste discharger.

These costs include those for the common waste treatment plants as well as the cost of industrial wastewater reduction practices, interceptor sewers, water quality surveillance, stream low-flow augmentation, and possible in-stream aeration.

Land Value Benefits. Bordering land value benefits at a given water quality may also be estimated for a given land use by comparing the per acre market value of shoreline property with the value of nearby nonshoreline property. These market values may be estimated by using local tax records and the tax equalization rates. The difference in these per acre values will then reflect the unit benefits or damages of the shoreline location. Values at a higher water quality may be estimated by applying this technique to a nearby lake and projecting the ratio of the shoreline to nonshoreline per acre values to the original lake.

In-Stream Water User Benefits. In-stream water uses include commercial fishing, barge and ship navigation, flood control, and hydroelectric power generation. The value of commercially caught fish may be taken as a benefit, whereas the other uses involve damages or negative benefits.

DETERMINING A UNIT CHARGE FOR WATER-RELATED RESOURCES

The relationship of treatment costs to benefits may be used to arrive at a unit charge for water-related resources. The following is a representative case.

In 1972 I proposed that a regional board be empowered to sell the assimilative capacity of a specific water resource to those dischargers using the resource. Such a board would require at least the following information:

1. Identity of all discharges

2. Quantities of discharges
3. Existing and desired pollution index (a measure of water quality)
4. Benefits of waste treatment
5. Assimilative capacity of the water resource

Using a method developed and proposed by our group previously,[9] one can then determine the present value of the pollution index of the water resource. The pollution index is developed for specific water uses when multiple items of water quality are considered. It is specific for one of three classification of water use: human (direct) contact, indirect contact, and remote contact. An overall pollution index can be developed as a weighted average of three specific classification indices, given above, the weight of each being the proportion of the three types of use of the watercourse. Essentially, the formulation is

$$
PI_{ij} = \sqrt{\frac{\text{Max}\left(\dfrac{C_i}{L_{ij}}\right)^2 + \text{Mean}\left(\dfrac{C_i}{L_{ij}}\right)^2}{2}}
$$

```
P I SUB ij = SQRT { { Max( C SUB i OVER L SUB ij )
SUP 2 + Mean ( C SUB i OVER L SUB ij ) SUP 2 } OVER 2}
```

where

C_i = multiple items of water qualities

L_{ij} = permissable levels of the respective items for a use j and

PI_{ij} = Pollution Index for the use j.

See footnote[9] for complete illustrative explanation.

---

[9] N. L. Nemerow, *Stream, Lake, Estuary, and Ocean Pollution* 2nd ed. (New York: Van Nostrand Reinhold, 1991) 272–286.

After itemizing all the annual benefits of uses of a water resource at its existing quality level, we can obtain two points on a curve, as shown in Figure 2.3 (illustrative of actual data collected on Onondaga Lake, Onondaga County, New York State).[10]

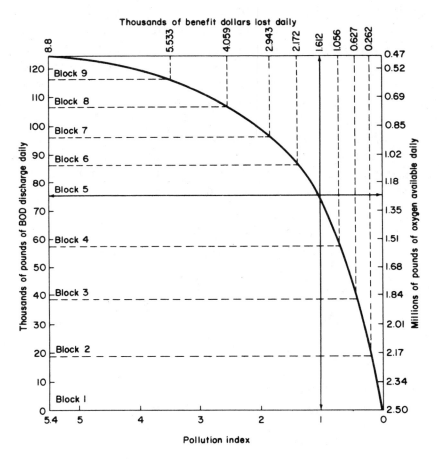

Figure 2.3. Pollution Index Measurement

[10]N. L. Nemerow, "Cost of Water Pollution Control," Paper presented at the National Symposium on Cost of Water Pollution Control, Raleigh, N. C., April 6–7, 1992, page 238.

These points are very significant: point A, the coordinates of which are 5.4, the computed present pollution index of the lake, and 1,230,000 lb BOD/day, the present daily BOD inflow into the lake. The other point, B, has coordinates of 0 for pollution index (representing absolutely no contaminants) and 0 pounds of BOD being discharged into the lake each day. Although no precise data are available between these two factual points, we can make some assumptions that will help define the shape of the curve. First, the curve should descend in BOD toward a resultant pollution index of zero, but probably not precisely in a linear fashion.

We also know that point A corresponds to $8,800 of benefits lost daily at the present discharge level and that point B corresponds to no loss in benefits when there is no discharge of wastes. A third point, C, can be approximated at the intersection of the two objectives, PI = 1.0 (PI = Pollution Index) and dissolved oxygen equal to 4 milligrams per liter. The curve can now be drawn with considerable certainty through the three points A, B, and C. Point C presumes that a direct relationship exists between the daily BOD discharge (76,000 lb) and dissolved oxygen available (1.2 million lb per day at 4 ppm). From this curve, we can proceed to make a firm decision on what water quality to maintain, based on constraints of dollar benefits and amount of oxygen, with a resultant allowable discharge of BOD. In our illustration (Figure 2.3), we decide to improve the PI to 1.0 from its current value of 5.4. This will result in a decrease in dollars of benefits lost from $8,800 per day to $1,612 per day (a savings of $7,188 per day). Treatment to a level of 76,000 lb BOD per day (37.4% reduction) or sale of BOD capacity resources up to a level of 76,000 lb per day are the chosen two alternatives open to a board. If sale of resources is chosen, four blocks of 19,000 lb BOD each can be sold at costs directly related to the dollar benefits lost for each incremental amount of BOD contamination allowed. For example, the third 19,000 block results in a benefit loss of $1,056,627 or $429 per day. Therefore, a potential and reasonable unit charge could be $429/19,000 lb, or 2.3 ¢/lb. Customers (water-polluting users) may choose to build their own waste treatment plants to eliminate the need for purchase. Note that each block of BOD capacity purchase becomes increasingly more expensive. Presumably, no more than 76,000 pounds would be sold unless a lower water quality was desired, and less would be sold if a higher quality was selected by the board. Customers will probably select the least-cost alternative to comply with allowable BOD discharge.

Purpose of Quantified Benefits

Quantified benefits serve two major purposes. First, they allow one to compare the total annual dollar benefit with a total annual expenditure, or cost of maintaining or achieving a particular water quality (pollution index). In the Onondaga Lake analysis, it was found that the annual benefits of the improved water quality (PI = 1.0) would be about $7.5 million greater than those of its existing water quality level (PI = 5.4). Annual expenditures of capital and interest payments to obtain this improved water quality have been estimated to be about the same, $3.2 million. Therefore, we can show that the quantifiable benefits alone would equal the annual costs. In addition, there would certainly be intangible benefits, ample to dictate a "cleanup" policy to the board.

Second, qualified benefits allow one to compute a unit charge for a pollution-carrying capacity resource. For example, in our illustration the sale of the third block of BOD at 2.3¢ per pound will decrease water quality from a pollution index of about 0.4 to about 0.75 (See Figure 2.3).

All pollution capacity consumers (BOD purchasers) are now in a position to select one of the two alternatives: (1) to purchase BOD capacity at 2.3 per pound or (2) to treat its own wastes at a lower cost. In deciding between the two, the consumer must consider both capital and operating costs. Typical capital costs for many organic industrial wastes range from $150 to $300 per pound of BOD per day. Operating costs for industrial waste treatment plants are relatively high and extremely variable, and are unavailable for public analysis. It is an unfortunate fact that many industrial plants themselves have not assessed the true operating costs for waste treatment. Municipal operating costs approximate $200 per mgd for secondary plants, or about 10¢ per pound of BOD.

Affordability

Can an industrial plant really afford waste treatment? Neither an industrial plant nor a governmental regulatory agency really knows the true answer to this question. If we discount as unrealistic, the response that a plant must treat its wastes regardless of the cost, then we are obliged to seek a reasonable reply. We must provide an objective and feasible method for ascertaining a treatment cost that a plant can afford and still remain competitive in its industrial category.

In a 1973 survey, detailed information was obtained by questionnaire from four of nine selected plants. These plants were chosen because of our previous knowledge of their cooperative participation in effective waste treatment while remaining highly competitive. The four dependable replies were evaluated and summarized in Table 2.3.

A study of the table revealed that two relationships of industrial economics have potential for application in water pollution abatement. The first is the ratio of dollars

Table 2.3 Summary of Cost Data of Four Plants

Plant No.	Type product	(Cost ratios based on product economics)					(Cost ratios based on waste flows)		(Cost ratios based on BOD removals)		(Cost ratios based on suspended solids removal)		Ratio cost based upon total plant asset
		Waste treatment cost / Dollar value added *(%)	Waste treatment cost / Dollar value of raw materials	Waste treatment cost / Dollar value added plus raw material (%)	Waste treatment cost / Dollar value of selling price (%)	Waste treatment cost / Dollar value of profit and overhead (%)	Waste treatment cost / Gal per day ($/gpd)	Waste treatment cost / 1000 gal waste ($/1000 treated gals)	Waste treatment cost / Pound BOD per day ($/lb BOD per day)	Waste treatment cost / Pound BOD ($/lb BOD)	Waste treatment cost / Pound SS removed in a day ($/lb SS/day)	Cost / Suspended solids removed ($/lb/SS)	Waste treatment cost / $ Plant asset (%)
1	Beans	1.11	2.16	0.795	0.51	1.64	.0236	0.374	10.30	0.1635	18.50	0.294	0.574
2	Tomatoes, peaches	1.08	—	—	—	—	.007	0.233	1.71	0.057	2.28	0.076	1.50
3	Chickens	1.096	0.159	0.139	0.122	0.89	.0484	0.186	1.08	0.00417	1.92	0.0074	0.925
4	Upper leather	0.42	0.245	0.155	0.101	0.29	0.1035	0.393	47.00	0.004	0.957	0.000329	1.04

*for meaningful relationship for water pollution abatement.

Source: N. L. Nemerow and C. Ganotis, "Benefit related expenditures for industrial wastewater treatment" Water and Sewage Works, Reference Number, p. R, 128, 1973.

of waste treatment cost to production or "value added." Three of the four plants reported a 1% (±0.1%) ratio. The fourth plant showed only a 0.42% ratio. Indications are that a plant may allocate a percentage of its production cost to waste treatment. A very preliminary rough approximation of this percentage showed approximately 1% as a fair value. However, these plants utilized very economical, as well as effective, waste treatment methods. The second ratio results are 18¢ to 39¢ cost of waste treatment for each 1,000 gallons of waste treated.

All costs of waste treatment include annual costs of amortized capital expenditures, maintenance, power, and chemicals. A 10-year useful life of equipment and a 7% interest rate were used to compute annual capital costs.

The most important question regarding the usefulness of either of these methods is that of determining whether a plant can, in reality, afford to spend 1% of its production cost, or even 24¢ per 1,000 gallons of waste, for treatment. The financial profitability as a result of business and professional management determines the ability of a plant to compete with others and still be able to provide adequate waste treatment.

An attempt was then made to compute significant values such as marginal income and profit: sales ratios. Total sales revenue of a plant minus the direct costs of the plant would be classified as marginal income. Direct costs include basic raw materials used to produce the product, variable labor costs, utilities used in the production of the product, and waste treatment costs. All other costs are fixed and include such items as management salaries, rent, mortgage and interest, loans, taxes, depreciation, waste treatment capital costs, and others. When fixed and direct costs are added and the sum is subtracted from sales revenue, one obtains the true profit income. The resulting ratios of marginal and profit income values to the sales revenues for each plant of a given industry can be compared in some mathematically valid manner to arrive at a comparative financial index, such as:

```
{ Marginal Income } OVER { Sales Revenue } + { Profit
Income } OVER { Sales Revenue } = Financial Index
```

$$\frac{\text{marginal income}}{\text{sales revenue}} + \frac{\text{profit income}}{\text{sales revenue}} = \text{financial index}$$

The higher the index, the more financially able the plant should be to afford waste treatment. When comparing indices of all plants of an industry, one can decide by use of a statistical value and the industry's agreement on which plants should be able to provide waste treatment and still remain economically competitive.

Unfortunately, although the method is sound, industry either does not know its real direct and fixed costs or will not disclose these to "outsiders." Without such information, the method is useless. Perhaps at some time in the future, if and when marginal income and profits of each separate plant are known and made public, this method will provide a means of decision making.

Determining a Firm's Financial Capability to Provide Waste Treatment

A method is being developed to establish a basis for determining both a reasonable amount of expenditure that could be spend for treatment, and for compelling the most financially able companies to initiate treatment. Financial potential is measured by the ratio of pollution abatement cost (to meet required environmental quality) to sales revenue for each firm, hereafter referred to as the sales index. For a firm with a sales revenue of $100 million annually and a treatment cost requirement of $5 million, the sales index should be .05. Indices can be calculated in a similar manner for all firms in the industry, placed in an array, and analyzed statistically. A hypothetical comparison of the indices of five firms in one industry is shown in Figure 2.4.

Indices vary from 2.5¢ to 11¢ per sales dollar. In our illustration, the firms in the industry arrived at the consensus that an index of more than 10¢ per sales dollar, would result in an economically impossible situation. Therefore, firm A must increase its sales dollars, internalize an increased profitability, or receive some sort of subsidy so as to be able to pay for pollution abatement. Its other alternative is to cease or alter its production. On the other hand, firm D, with the lowest index of 2.5¢, would be most financially able to treat its waste first (with a resulting change in its index). A new redistribution of the indices is then computed (See Stage II in Figure 2.4) with a mean index increasing toward the cutoff point of 10¢, and firm E, with an index of 4¢, next in line for treatment.

The Sales Index Method was applied to the pulp and paper industry, the data of which were published. The sales values and treatment costs and sales indices are shown in Table 2.4. An industrial average of $30.68 million per company would be needed to meet the requirements.

In addition, it would cost an average of 4.34¢ per sales dollar (Sales Index) to attain this level of waste treatment. Descriptive statistics of sales index data showed widespread differences among firms, as indicated by a range of 0.3¢ to 11.75¢ per sales dollar. The relatively high standard deviation implies that pollution abatement activity in this industry varies significantly among firms. A cutoff point (upper limit of Sales Index) would then have to be determined by a caucus of the 24 companies. Presumably, Company 16 (Sales Index 3.0) is in the best financial position to install necessary waste treatment facilities, whereas Company 10 (Sales Index 117.5) may already be financially incapable of affording waste treatment.

Hypothetical index calculations:

Firms	A	B	C	D	E
1. Proposed treatment cost (85% removal)	$110,000	$120,000	$30,000	$20,000	$18,000
2. Gross sales:	$1,000,000	$1,300,000	$600,000	$800,000	$450,000
Index $(^{(1)}/_{(2)})$ $/per sales $	$.11	$.093	$.05	$.025	$.04

Assume: Prior Industrial Agreement has been made that it is feasible to spend up to 10¢ per sales dollar before financial crisis occurs.

STAGE 1 Average Index: 5.3/¢ sale, Firm D potentially will be most able to treat first.

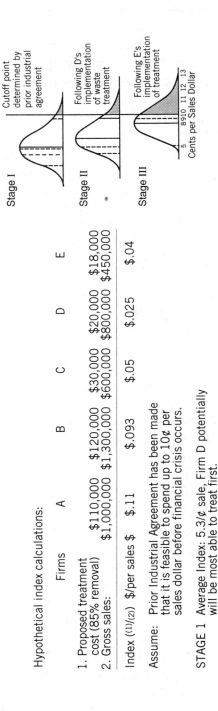

Stage I — Cutoff point determined by prior industrial agreement

Stage II — Following D's implementation of waste treatment

Stage III — Following E's implementation of treatment

5 8 9 10 11 12 13
Cents per Sales Dollar

Figure 2.4. Statistical Treatment of Hypothetical Data of Treatment Costs and Sales

Table 2.4 Pulp and Paper Industry Treatment Costs, Sales Revenue (1970), and Sales Indices (Computed)*

Industrial Corporation	(A) Treatment Cost ($ Millions)	(B) 1969 Sales in ($ Millions)	Sales Index $\frac{A}{B} \times 10^{-3}$
1. Amer. Can	12.5	1723.7	7.3
2. Boise-Cascade	42.1-50.1	1726.0	26.8
3. Consolidated Paper	9.29	127.7	7.3
4. Cont. Can	16.5	1780	9.3
5. Crown Zeller	58.7	919.3	64.0
6. Diamond Int.	8.6	498.1	17.3
7. Fibreboard	4.8	181.8	26.4
8. Ga. Pac.	23.6	1160.2	20.3
9. Gt. Northern Nek.	25.2-34.2	340.7	86.5
10. Hammermill	41.5	353.3	117.5
11. Hoerner-Wald.	17.5	237.3	73.7
12. Int. Paper	101	1777.3	57.0
13. Kimb-Clark	13.5	834.7	16.2
14. Marcor	28.3	2500.7	11.3
15. Mead	26	1038	15.1
16. Owens Ill	3.9	1294.4	3.0
17. Potlach	28-37	337.1	96.5
18. Riegal	11.5	184.0	62.5
19. St. Regis	59.8	867.8	69.0
20. Scott	78.8	731.5	107.5
21. Union Camp.	9.0	449.5	20.1
22. U. S. Plywood	13.2-21.2	1455.5	11.8
23. Westvaco	36.5	419.6	87.0
24. Weyerhaeuser	32.5	1239.2	26.2

*Capital and Estimated Operating Cost to reach desired treatment level as required by federal and state governments (average costs used in computation of Sales Index).

Source: N. L. Nemerow and C. Ganotis, "Benefit related expenditures for industrial wastewater treatment" *Water and Sewage Works*, Reference Number, p. R, 128, 1973.

Although we know that Company 16 is potentially capable of additional expenditure for waste treatment, and that Company 10 is potentially incapable of the same expenditure, this method still does not assess a company's actual financial ability to pay. It gives the companies and regulatory agencies a valid indication, however, that one company should be able to afford treatment costs and another will not be able to afford the same costs. These conclusions in themselves should reveal facts to both the companies and to society which, heretofore, had been undisclosed. The findings should provide both with directives for future action leading to elimination of excessive pollution.

W. W. Doerr, with a basic education in economics and advanced education in chemical engineering, also advocates—after identifying the problem and techniques to reduce pollution—establishing the associated capital and operating costs. He proposes generating a "cost per unit of pollutant." Doerr says that we can obtain data on capital and operating costs from EPA clearinghouse sources, and electronic bulletin boards, state authorities, vendors, industry literature, corporate data, and symposia.

He specifies that capital costs include new equipment, with salvage costs credits of replaced equipment considered. Operating costs should include direct costs of utilities, chemicals, and labor. These latter costs should consider the waste hauling or treatment required, *or avoided*, and changes in future disposal liabilities, improved marketability of manufactured products *that reveals improvements resulting from environmental awareness*, and changes in labor costs associated with waste hauling, handling, or disposal. Doerr recognizes that industry does benefit from an improved environmental image. For example, the Dreyfus family of funds includes the Dreyfus Third Century Fund, a collection of environmentally conscious industrial firms.

Doerr recommends standardizing the costs analysis to an appropriate method of analysis such as *net present value, discounted cash flow analysis, rate of return, net capital costs, and annual net operating costs or revenues.* He also suggests using the unit cost factor of "dollars per pound of pollutant reduced" to compare technical options for pollution reduction.[11]

---

[11]W. W. Doerr, "Plan for the Future with Pollution Prevention," *Chemical Engineering Progress* (January 1993): 24–29.

CHAPTER 3

WASTE MINIMIZATION BY REUSE AND RECOVERY

FOUNDATION

The first stage of an industrial plant's waste minimization program must be a realization that reducing waste is not only an external environmental asset or enhancement, but also an economic benefit from a production standpoint. The goals of such an effort should be to minimize adverse environmental impacts and, at the same time, reduce production costs—both can be attained through a program of reuse and recovery.

THE PROCESSES

The obvious first approach in a plant's reuse and recovery program centers on recovering and reusing its own waters within its own facility. This can be done best through understanding the production processes in detail and then supplementing this knowledge by making mass balances (*input and output quantities*) about each process. Continuous review and study of these mass balances will enable the environmental engineer to contemplate changes in processes to decrease wastes, to ferret out potential sources of wastes that may be recovered and reused, and to predict how significant alterations will affect waste production.

The system then calls for a program of sampling and analysis of all major wastes. Analysis must include *volumes* and *frequency* of waste discharges, *physical* nature of wastes (solid, liquid, or gaseous), *chemical state* of waste components (such as lead in solid wastes, polychlorinated biphenyl (PCB) compounds present in liquid wastes, and acids or fluorine present in gaseous effluents. An analysis program must be derived specifically for each industrial plant, inasmuch as the wastes discharged depend not only on the type of industry, but also on raw materials and method of processing used. Similar plants producing the same manufactured product may discharge wastes differing in character and amount. A good idea of the wastes that can be expected from various industrial operations may obtained by referring to page 21 of *Industrial and Hazardous Wastes*.[1]

However, no reference book by itself will reveal exact waste quantities nor even precise pollution characteristics.

SOLID WASTE RECYCLING

Once the quantities and characteristics of each type of waste are ascertained through reference to the literature and sampling and analyses, they can be evaluated for potential reuse within the plant. The environmental engineer must now begin an assessment of both the water used in production and the raw materials used in manufacturing. Assessment should include allowable tolerances of contaminants in the raw process and the raw materials used to manufacture the finished product. For example, a bakery's raw water should contain less chloride and carbonate than sulfate, because the presence of sulfate often enhances the final bread product; or an insulating papermill's raw processing water could be unacceptable with an excess of chlorides, which increases the paper's conductivity, while the amount of carbonates could be higher. Likewise, a Portland cement plant's raw material can tolerate 3% by weight of calcium sulfate in addition to calcium carbonate to manufacture calcium oxide cement, but more than 0.6% of these alkalies may be intolerable for the final product.

---

[1]N. L. Nemerow and A. Dasgupta *Industrial and Hazardous Wastes* (New York: Van Nostrand Reinhold, 1991), 21.

In some cases, actual pilot experimentation must be used to determine the acceptability of reusing a wastewater or solid residue within its own plant. To this end, it is important that all industrial plant personnel maintain an open mind to reuse. Most plant superintendents have a real sense of ownership regarding their manufacturing processes; their cooperation must be solicited diplomatically, not dictated in an argumentative manner. Reuse of as much process waste and solid residue as possible within the plant is the procedure most beneficial to all parties concerned in a waste minimization program. It can reduce process water requirements and, hence, the costs of production. Reuse of solid wastes can also reduce or eliminate environmental costs as well.

In its April 1988 publication EPA/600/2-88-025, hereafter referred to as EPA (1988), the Environmental Protection Agency described (in graphical form) the techniques of waste minimization. They are reproduced here in Figures 3.1 and 3.2. Although it should be obvious to our readers that waste minimization by industry is good business, some reasons highlighted by the EPA are given here as support rationale. These are mainly concerned with economics and include the following: reduced waste disposal costs (because less waste results), reduced raw material costs (because manufacturing processes more efficiently use and reuse these materials), reduced damage costs (because lawsuits and insurance charges are eliminated), and improved public relations (because both the local and national public image of the industry is elevated as a result of environmental improvement). In addition, the paperwork associated with national, state, and local permitting regulations is greatly reduced, providing savings of both time and money.

EPA (1988) separates waste minimization into two categories: source reduction and recycling, both within and

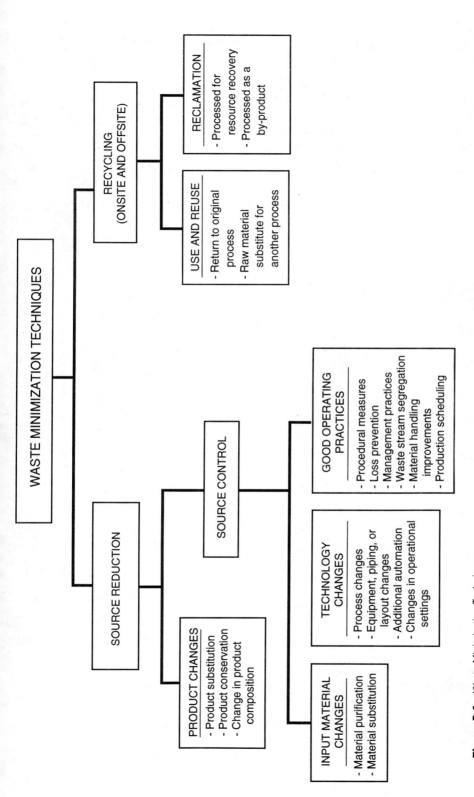

Figure 3.1. Waste Minimization Techniques
Source: EPA (1988).

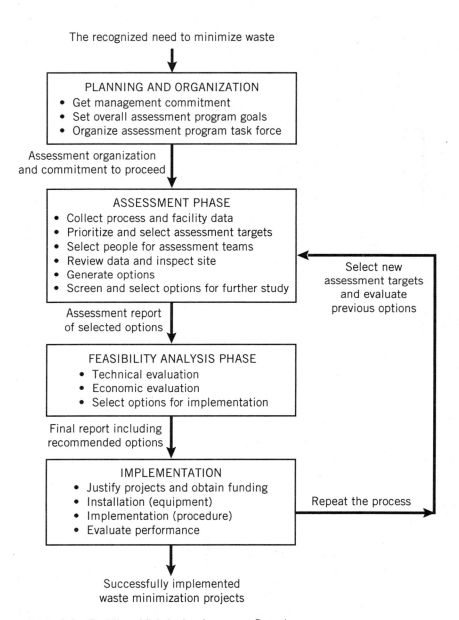

The recognized need to minimize waste

PLANNING AND ORGANIZATION
- Get management commitment
- Set overall assessment program goals
- Organize assessment program task force

Assessment organization
and commitment to proceed

ASSESSMENT PHASE
- Collect process and facility data
- Prioritize and select assessment targets
- Select people for assessment teams
- Review data and inspect site
- Generate options
- Screen and select options for further study

Select new
assessment targets
and evaluate
previous options

Assessment report
of selected options

FEASIBILITY ANALYSIS PHASE
- Technical evaluation
- Economic evaluation
- Select options for implementation

Final report including
recommended options

IMPLEMENTATION
- Justify projects and obtain funding
- Installation (equipment)
- Implementation (procedure)
- Evaluate performance

Repeat the process

Successfully implemented
waste minimization projects

Figure 3.2. The Waste Minimization Assessment Procedure
Source: EPA (1988).

outside the plant, and recovery and exernal sale. The latter method is discussed in detail in Chapter 4; source reduction and recycling within the plant are described in the following pages of this chapter.

EPA (1988) also proposes that a waste minimization program proceed through sequential operations of planning and organizing, assessment, feasibility analysis, and implementation. If these four steps are followed carefully and effectively, the discharge of industrial wastes can be reduced to a minimum.

In planning and organizing, the director designate of waste minimization brings the company's proper functional groups together to plan for waste reduction. The personnel of these groups selected by the director designate must be respected sufficiently to obtain management's approval and commitment to waste the reduction program. The group selected should set its own objective criteria and staff subgroups for special tasks. One obvious objective should be waste reduction with resulting benefits to industry that exceed the cost of implementation. Another objective should be the involvement of all workers (especially those employed in the waste production process) in programs of waste minimization. Often, these workers must be given incentives to participate in such a plan. Incentives may include monetary compensation, special plaques or certificates, or publicly presented awards. At this point, any of several impediments within the company can pose a major problem in achieving success. EPA (1988) lists nine of these barriers. Most are related to attitudes that have been ingrained in industry personnel for a long time, for example: *It will cost us more money. It will deteriorate our prime product. The trouble will not be worthwhile to anyone personally. It will likely lead to some form of internal conflict within the company.* Each department or division of the company has inherited a built-in adversity to reducing waste. We list here seven departments and some objections to reducing waste, which must be overcome by education.

Department	Objection
Finance	We will be unable to afford the expense.
Purchasing	Change in raw materials we purchase may affect the production flow.
Marketing	Product changes will make selling the new product more difficult.
Quality monitoring	New guidelines and/or more work may be necessary.
General Engineering	Space, services, or manpower may be inadequate.
Environmental	Existing waste treatment may not be effective, and changes mean (at the least) that new regulations will be imposed on the plant.
Production	Changes will decrease the output, even though temporarily.

All of these objections are caused by fears that more work may be required with no guarantee that it will be justified. Education of personnel and strong leadership by administrators are needed to overcome these objections.

To ferret out all the options available to attain zero pollution, one needs to analyze information about the processes used in the industry: the exact amounts and types of wastes produced, evaluation and segregation of hazardous types, materials used and lost in processing, and general in-plant procedures for preventing losses in materials of all types.

Table 3.1 lists details of information needed at this point, as provided in EPA (1988).

Table 3.1   Facility Information for Waste Management Assessments

Design Information

Process flow diagrams
Material and heat balances (both design balances and actual balances) for
      Production processes
      Pollution control processes
Operating manuals and process descriptions
Equipment lists
Equipment specifications and data sheets
Piping and instrument diagrams
Plot and elevation plans
Equipment layouts and work flow diagrams

Environmental Information

Hazardous waste manifests
Emission inventories
Biennial hazardous waste reports
Waste analyses
Environmental audit reports
Permits and/or permit applications

Raw Material/Production Information

Product composition and batch sheets
Material application diagrams
Material safety data sheets
Product and raw material inventory records
Operator data logs
Operating procedures
Production schedules

Economic Information

Waste treatment and disposal costs
Product, utility, and raw material costs
Operating and maintenance costs
Departmental cost accounting reports

Other Information

Company environmental policy statements
Standard procedures
Organization charts

In collecting information, material balances to reveal potentially ineffective operations are suggested as important for another reason. Material balances are often needed to comply with Section 313 of SARA (Superfund and Reauthorization Act of 1986) in establishing emission inventories for specific toxic chemicals. EPA's Office of Toxic Substances (OTS) has prepared a guidance manual entitled *Estimating Releases and Waste Treatment Efficiencies for the Toxic Chemicals Inventory Form* (EPA 560/4-88-02, EPA, 1988).

The options available to industry include reducing wastes during processing, recycling, and, finally, treatment of wastes. The latter should be considered only after fully utilizing and/or discounting waste reduction and recycling. Good operating practices generally result in reduction or even minimization of wastes during processing. Most practices are controlled by plant personnel and changes can be implemented readily by management. In addition to minimizing wastes through planned programs, good operation can be accomplished by the following: (1) improving material handling and inventory practices; (2) loss prevention by eliminating leaks and spills; (3) waste segregation by separating treatable from nontreatable wastes; (4) improving cost accounting methods by departmentalizing waste costs; and (5) better production scheduling to prevent batch operations, which require frequent cleanups.

During the assessment, all of the waste management options are elucidated. These must be evaluated from both technical and economical standpoints. In deciding on the technical possibilities of an option, the manager must ascertain whether the practice will work in the plant. Each industrial plant is unique and warrants separate consideration. Certain facilities such as space and utilities, may be available at one plant and not at another. In addition, each plant is obligated to manufacture a product to specific requirements. The quality of the product often dictates whether a change to benefit waste management is possible.

We assess the economic potential of a change in waste management on the final basis of capital and operating cost changes. If a plant assesses the economic benefits of a waste minimization change, it must include the specific benefit of avoiding the consequences of pollution. (Refer to Chapter 2 for a more detailed economic evaluation.) If a specific piece of equipment is required to augment the waste minimization technique, one may use the payback period or return-on-investment method to compute the real cost. The payback period in years is figured by dividing the cost of the equipment by the annual savings in operating costs.

EPA (1988) adopted a typical capital investment detail for a large venture from Perry, *Chemical Engineers Handbook* (1985), and Peters and Timmerhaus, *Plant Design and Economics for Chemical Engineers* (1980). It presents a detailed study of the capital investment for a typical large waste minimization project, including capital costs and working capital.

Direct Capital Costs

1. Site development
2. Process equipment
3. Materials
   Piping
   Insulation and painting
   Instrumentation and controls
   Buildings and structures
4. Connection to existing utilities and services
5. Other nonprocess equipment
6. Construction/installation

Indirect Capital Costs

1. In-house engineering, procurement, and other home office costs.
2. Outside engineering, design, and consulting services
3. Permitting costs
4. Contractor's fees
5. Start-up costs
6. Training costs
7. Contingency
8. Interest accrued during construction

A summation of the above Direct and Indirect Capital Costs yields the Total Fixed Capital Costs.

Total Fixed Capital Costs = Direct Plus Indirect Costs

Working Capital

1. Raw material inventory.
2. Finished product inventory.
3. Materials and supplies.

The sum of 1-3 above equal the Total Working Capital.

Total Working Capital

Finally we can sum the Total Fixed Capital Costs and Total Working Capital to obtain the Total Capital Investment

$$\text{Total Capital Investment} = \text{Total Fixed Capital Costs} + \text{Total Working Capital}$$

EPA (1988) points out that by using waste minimization techniques an industry obtains the advantage of eliminating permitting costs, which are increasing with all forms of environmental regulations.

The lower the waste disposal costs, the better any industrial waste minimization program is operating. The ultimate objective of such programs is to eliminate completely all so-called waste disposal costs. When this is accomplished, profits (and product salability) will be maximized and, at the same time, the environment will be protected.

EPA (1988) lists the types of waste minimization production changes and their associated savings in costs. These are shown in Table 3.2. EPA (1988) also gives typical costs for disposal of hazardous wastes offsite (Table 3.3).

Without an effective waste minimization program, an industrial plant may face an unpredictable risk of fines,

Table 3.2 Operating Costs and Savings Associated with Waste Management Projects

Reduced Waste Management Costs

This includes reductions in costs for:
    Offsite treatment, storage, and disposal fees
    State fees and taxes on hazardous waste generators
    Transportation costs
    Onsite treatment, storage, and handling costs
    Permitting, reporting, and recordkeeping costs

Input Material Cost Savings

An option that reduces waste usually decreases the demand for input materials.

Insurance and Liability Savings

A WM option may be significant enough to reduce a company's insurance payments. It may also lower a company's potential liability associated with remedial cleanup of Treatment, Storage, and Disposal Fees and workplace safety. (The magnitude of liability savings is difficult to determine).

Changes in Costs Associated with Quality

A WM option may have a positive or negative effect on product quality. This could result in higher (or lower) costs for rework, scrap, or quality control functions.

(continued)

Table 3.2 (continued)

Changes in Utilities Costs

Utilities costs may increase or decrease. This includes steam, electricity, process and cooling water, plant air, refrigeration, and inert gas.

Changes in Operating and Maintenance Labor, Burden, and Benefits

An option may either increase or decrease labor requirements. This may be reflected in changes in overtime hours or in changes in the number of employees. When direct labor costs change, then the burden and benefit costs will change. In large projects, supervision costs will also change.

Changes in Operating and Maintenance (O & M) Supplies

An option may increase or decrease the use of O&M supplies.

Changes in Overhead Costs

Large WM projects may affect a facility's overhead costs.

Changes in Revenues Resulting from Increased (of Decreased) Production

An option may cause an increase in the productivity of a unit. This will result in a change in revenues. (Note that operating costs may also change accordingly.)

Increased Revenues Derived from By-products

A WM option may produce a by-product that can be sold to a recycler or sold to another company as a raw material. This will increase the company's revenues.

lawsuits, and criminal penalties for lack of compliance with environmental regulations. EPA (1988) suggests that industry include these "risk" costs in the payback (return of capital) period. A higher period, say 10 years, would account for some of the unpredictable costs given above. EPA suggests that by using this relaxing of economic acceptability, such a procedure is often preferable to a company's financial officers.

As corroboration of this policy, EPA administrator William Reilly in 1988 stated, "H.R. 1457, the Waste Reduction Act, is a big step in the right direction; the bill's

**Table 3.3   Typical Costs of Offsite Industrial
Waste Management**[a]

Disposal
   Drummed hazardous waste[b]

Solids	$75 to $110 per drum
Liquids	$65 to $120 per drum
Bulk waste	
Solids	$120 per cubic yard
Liquids	$0.60 to $2.30 per gallon
Lab packs	$110 per drum
Analysis (at disposal site)	$200 to $300
Transportation	$65 to $85 per hour @ 45 miles
	per hour (round trip)

[a]Does not include internal costs, such as taxes and fees, and labor for manifest preparation, storage, handling, and recordkeeping.
[b]Based on 55-gallon drums. These prices are for larger quantities of drummed wastes. Disposal of a small number of drums can be up to four times higher per drum.

emphasis on preventing pollution at the source rather than 'end of the pipe' control is appropriate and welcomed."[2] Techniques for waste minimization include *source reduction* and *recycling*. Details of these techniques are shown schematically in Figure 3.1 Source reduction is inhibited primarily by a lack of awareness or lack of information, rather than technical constraints, economic costs, or governmental prohibitions.[2] However, E. A. Flores believes that the obvious current incentives for waste minimization are impeded by institutional, economic, technical, and behavioral factors.[3]

Institutional impediments are mainly EPA's recycling regulations. Economic problems are blamed on accounting methods that fail to recognize indirect savings or intangibles. Technical roadblocks usually involve lack of exper-

---

[2]P. N. Cheremisinoff and J. A. King, "Waste Minimization," *Pollution Engineering* (March 1991): 64.
[3]E. A. Flores, "Impediments to Haz Waste Minimization," *Pollution Engineering* (March 1991): 76.

tise to make necessary operational changes. Fear of the unknown and lack of positive evidence of total benefits of making these environmental changes are the major factors leading to behavioral impediments.

Recycling potential depends on where you live and what you are attempting to recycle and sell. Typical revenues for newspapers, aluminum cans, clear glass, plastic, and white paper, in vertical order, are given in dollars per ton in Figure 3.3.

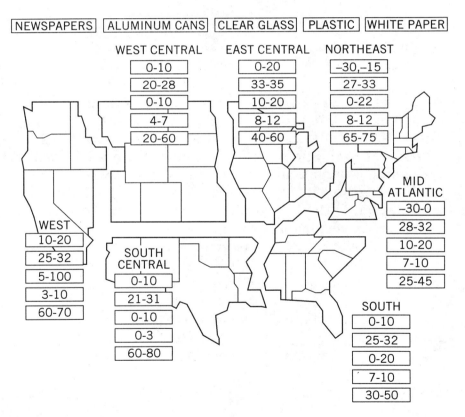

Figure 3.3. Recycling Revenues
Source: T. Arrandale, Governing Guide—Recycling: Getting Down to Business, *Governing* (August 1991): 39.

Recycling is enhanced by the positive attitude and active participation of the original manufacturer in reusing the material. T. Arrandale provides ample evidence of this in the aluminum, plastic, and steel industries. Incentive programs in the lead acid battery and automotive oil industries also influence the amount of recycling. In quoting Chaz Miller of the Glass Packaging Institute, Arrandale confirms my contention that "recyclables are simply a raw material, nothing more and nothing less."[4]

For the benefit of the reader, in Figure 3.4 we have listed the names, addresses, and telephone numbers of some major organizations in the recycling business. In Dade County, Florida, consumers will be able to take aluminum cans to a "reverse vending machine," which crunches the cans and spits out change. The Reynolds Aluminum Company estimates that there will be 10,000 of these grinding sites.

Figure 3.4. *Recycling Resources*

With producers buying into the recycling movement, their associations are prime sources of information about how, where, and how much in regard to recycling the particular materials they produce. Here are some of the main organizations you can call for more information:

The Aluminum Association
900 19th St. N.W.
Washington, DC 20006
(202) 862-5161

American Paper Institute
1250 Connecticut Ave. N.W.
Washington, DC 20036
(202) 463-2420

---

[4]T. Arrandale, "Governing Guide—Recycling: Getting Down to Business," *Governing* (August 1991): 39.

Figure 3.4. (continued)

The Council for Solid Waste Solutions (plastics)
1275 K St. N.W., Suite 400
Washington, DC 20005
1-800-2HELP90

Glass Packaging Institute
1801 K St. N.W., Suite 1105-L
Washington, DC 20006
(202) 887-4850

National Solid Wastes Management Association
Waste Recyclers Council
1730 Rhode Island Ave. N.W., Suite 1000
Washington, DC 20036
(202) 659-4613

National Tire Dealers and Retreaters Association
1250 I St. N.W. Suite 400
Washington, DC 20005
(202) 789-2300

Plastics Recycling Foundation
1275 K St. N.W., Suite 400
Washington, DC 20005
(202) 371-5200

Steel Can Recycling Institute
Foster Plaza X
680 Anderson Drive
Pittsburgh, PA 15220
(800) 876-SCRI
(412) 922-2772

EPA - Used Oil Coordinator
Hail code OS 301
U.S.E.P.A. 401 M Street S.W.
Washington, DC 20460.

nationwide to enhance consumer recycling. After flattening, the cans are weighed (about 30 cans to a pound), the machine dispenses the proper payment, and blows the cans into a storage bin for subsequent collection and delivery to Reynolds.

Oil companies have initiated used-oil recycling by consumers through selected service stations. For locations and details of other recycling resources, the reader is directed to the EPA reference shown in Figure 3.4.

One means of waste minimization through recycling is by source substitution. Whenever it is possible to substitute another material for one generating environmental hazards and costs resulting from its use and disposal, the practice is desirable. For example, mercury minimization is recommended because this material is hazardous to our environment. In late 1990 in Broward County, Florida, the municipal solid waste (MSW) management began diverting mercury-containing wastes before they reached treatment facilities and landfills.[5] Using a source substitution method, the county diverts nearly one ton of mercury annually from just one point source: mercury medical batteries. Battery usage per year per hospital ranged from 100 to nearly 6,000, depending on the number of beds and occupancy rates. Almost without exception, area hospitals were disposing of mercury and nickel-cadmium medical batteries in regular or bio hazardous waste containers. The most popular option with respect to 8.4-volt mercury medical batteries (widely used to power portable cardiac monitors) was to substitute zinc–air batteries. These batteries are usually more expensive, but last longer. Stocks of mercury batteries on hand were either depleted or returned to the original supplier. Each 8.4-volt battery weighs about 1.8 ounces, of which almost 28% is mercury. Broward County found that its method of battery management was "a highly cost-effective way of diverting significant quantities of mercury from the municipal solid waste."

---

[5] J. L. Price, "Managing Mercury Battery Wastes Through Source Substitution," *Municipal Solid Waste Management* (January-February 1992) p: 16.

In fact, Rayovac Corporation reported that annual consumption of mercury in batteries declined by 90% in the period 1984–1990.[6] Even greater reduction occurred in alkaline manganese batteries. Today, all alkaline manganese batteries produced in the United States (with the exception of button cells) have mercury concentrations below 0.025%, as compared with its original concentration of 1%. At Rayovac, these dramatic reductions were the result of both "an absolute reduction in weight and cubic volume of wastes and a reclassification of waste from hazardous to nonhazardous."

Mercury also occurs in U.S. shipboard wastes. The navy's program objective is to minimize hazardous wastes by generation-rate reduction, substitution of nonhazardous for hazardous materials, and waste treatment. It admits that mercury is one of the most hazardous materials carried on board U.S. navy ships. One source of mercury is the mercury nitrate used as a reagent with chloride in titrations to determine the chloride content of boiler feedwater. The resulting wastes contain an average of 30 ppm. of mercury. The navy found that 9,870 gallons of HgCl waste could be treated by only 17 full-size cartridges, each containing 2 liters of GT-73 resin. The fees to the navy levied by three ports (U.S.) varied from $1.75 to $3.23 per gallon of mercury waste, or $32,000, based on 1988 costs. On the other hand, the navy estimates a reduction in cost to $29 for turning in the cartridges.[7]

In 1988, Holland and McCartney reported on DuPont's program for waste minimization, especially on reuse and

[6] R. L. Balfour and T. J. Anderson, "Source Reduction and Waste Minimization at Rayovac Corporation." Paper presented at the Third International Seminar on Battery Waste Management, Deerfield Beach, Florida, November 4–6, 1991.

[7] Craig S. Alig and M. Bogucki, "Minimization of Shipboard Mercury Wastes," *Journal of the Air Pollution Control Association* 38, No. 9 (September 1988): 1174–1177.

Table 3.4   Typical Column Headers in Computer Printouts

Production Area	Waste Description	Hazardous Classification	Quantity Generated M lb/yr	Management Disposal Cost $M/yr	Minimization Method
VI023	Organic acid	Flammable	2	5.5	Recycle
NR126	Polymers	Caustic	50	25	Sale
GA462	Spent catalyst	Acidic	10	42	Reuse
ME621	Lab solvent	Ignitable	0.5	1	Fuel
BU215	Acid catalyst	Corrosive	40	16	Administrative control

Source: G. J. Holland and R.F. McCartney, "Waste Reduction in the Chemical Industry," *Journal of the Air Pollution Control Association* 38, No. 2 (February 1988), pp. 174–178.

recovery.[8] They developed a computer printout for five classes of recoverable wastes, which includes means of minimization (see Table 3.4).

They summarized their paper by concluding that "reducing waste" to remain competitive has been an important ingredient in successful businesses in the past. In the future, it will be absolutely essential.

J. E. Leeman of Conoco Petroleum Corporation believes other incentives besides competition is driving the American petroleum industry to recycle and reuse materials. These other incentives include land disposal bans, the cost of commercial disposal, the difficulty in obtaining Part B permits, the lack of storage capacity, and the inherent headaches involved with hazardous wastes.[9]

In the metal finishing industry, G. E. Hunt cites five processes utilized in recovering wastes from this industry (see Table 3.5). As expected, these technologies have advantages and disadvantages that must be considered when making decisions on reuse of materials.

---

[8] G. J. Holland and R. F. McCartney, "Waste Reduction in the Chemical Industry," *Journal of the Air Pollution Control Association* 38, No. 2 (February 1988): 174–178.

[9] J. E. Leeman, "Waste Minimization in the Petroleum Industry," *Journal of the Air Pollution Control Association* 38, No. 6 (June 1988): 814–823.

Table 3.5   Summary of Recovery Technologies

Technique	Advantages	Disadvantages
Electrodialysis	• Achieves higher concentration than reverse osmosis or ion exchange • Energy efficient • Organics not concentrated • Inorganic salts transport at different rates minimizes of unwanted inorganics return	• Feed must be filtered • Membrane sensitive to flow distribution, pH and suspended solids • Equipment uses multi-cell stacks, incurs leakage • Chemical adjustment of recovered material • Membrane life uncertain
Electrolytic Metal Recovery	• Recovers only metals • Results in salable, nonhazardous product • Energy efficient • Low maintenance	• Solution concentration must be monitored • Fumes may form and may require hood scrubbing system • Solution heating encouraged to maximize efficiency
Evaporators	• Established and proven technology, very reliable • Simple to operate • Widely applicable • Can exceed bath concentration	• Some units are energy intensive • Multi-stage countercurrent rinsing essential • Returns bath and impurities • Additional treatment may be needed to control impurities • May require pH control
Ion Exchange	• Low energy demands • Handles dilute feed • Returns metal as metal salt solution	• Requires tight operation and maintenance—equipment complex • Limited concentration ability • May require evaporation to increase concentration • Excess regenerate required • Feed concentration must be closely monitored
Reverse Osmosis	• Achieves modest concentration • Small floor space requirement • Less energy intensive than evaporation	• Limited concentration range of operation • Fouling of membranes due to feeds high in suspended solids; feed filtration essential • Membrane sensitive to pH • Some materials fractionally rejected • May require further concentration

Source: G. E. Hunt, "Waste Reduction in the Metal Finishing Industry," *Journal of the Air Pollution Control Association* 38, No. 5 (May 1988): 672-680.

J. W. Tillman surveyed 10 U.S. industries for recent achievements in source reduction and recycling as a means of complying with the directives of CERCLA (Comprehensive Environmental Response, Compensation, and Liabilities Act) or the "Superfund." His study, done for EPA, covered the following industries: metals fabrication, manufacturing of machinery (nonelectric), lumber products, electronics, textiles, petroleum, food products, chemical products, printing and publishing, and transportation-related industries. Two typical plants were covered in each of the 10 industrial categories. They reported the following successes in their recovery efforts: (1) cost savings by reducing waste treatment and disposal costs, raw material purchases, and other operating costs; (2) achievement of state and national waste management policy goals; (3) reduction of potential environmental liabilities; (4) protection of public health and worker health and safety; and (5) protection of the environment.[10]

The reader is directed to this document as source of details of reduction and recycling in these specific industries.

P. Mizsey, a Budapest, Hungary, chemical engineer acknowledges that waste minimization in the chemical industry has shifted from the end-of-pipe treatment to eliminating wastes at the source, in addition to environmentally sound internal recycling. He proposes that "the main goal is to utilize the waste of a plant in another plant; that is, to design closed-cycle processing.[11] I compliment Mizsey on this statement, for I have been advocating the practice since 1977. See Chapter 5 for a detailed

---

[10]J. W. Tillman, "Achievements in Source Reduction and Recycling for Ten Industries in the united States," EPA Contract No. 68-C8-0062 (Cincinnati, Ohio: EPA/600/2-91/051, September 1991).
[11]P. Mizsey, "Waste Reduction in the Chemical Industry: A Two-Level Problem," *Journal of Hazardous Materials* 37 (1994). 1–13. (Elsevier Science B.V.)

discussion of examples of complexes with two or more industries.

Peters, Daniels, and Wolsky believe that waste minimization in the chemical industry offers significant opportunities because of the large amount of materials and energy used. They suggest waste reduction techniques such as (1) improvements in process selectivity and/or conversion, (2) the ability to operate at lower temperatures and/or pressures, (3) processes requiring fewer steps, feedstocks with fewer inherent by-products, (4) more efficient equipment design, products and/or catalysts with longer lives, (5) more efficient unit operation, (6) innovative process integration, (7) avoidance of heat degradation of reaction products, (8) new uses for otherwise valueless by-products, and (9) elimination of leaks and fugitive emissions.[12]

Minimizing the amount of solvents discharged in industrial wastes is a growing area of concentration for many manufacturers. G. M. Carlton describes four such industries: semiconductor, metals, paints, and aerospace. He found that for large waste volumes air stripping was an effective, flexible method of treatment of aqueous waste streams containing solvent wastes. He computed costs of $0.19 per 1000 gallons treated with vapor phase control.[13]

Bechtel researchers also studied minimizing solvent wastes.[14] After investigating innovative technologies, they found that the criteria for evaluating them included technical suitability, ease of implementation, cost, and envi-

---

[12]R. W. Peters, E. J. Daniels, and A. M. Wolsky, "Research Agenda for Waste Minimization," *Water Science Technology* 25, No. 3 (1992) 93–100:

[13]G. M. Carlton, Solvent Waste Reduction Alternatives Symposia "Utilizing Air Stripping Technology for Pretreatment of Solvent Waste." I.C.F. Consulting Assoc. Los Angeles, Cal. Oct. 20, 1986, pages 14–32.

[14]*Waste Minimization Study for the Lawrence Livermore National Laboratory: Solvent Wastes* (July 31, 1987), Bechtel National Inc. P.O. Box 3965, San Francisco, CA 94119.

ronmental concerns. They determined from actual mini-mization cases that the payback period for all cases was less than four years, with more than half reporting a one-year payback. In a typical case, used solvents are recycled in a batch-operated distillation column. Stored used solvent is treated by separation to remove solids and other contaminants. Separation may be achieved by centrifug-ing, filtering, decanting, or some combination of the three. Solids are land buried—if found to be nonhazardous. The separated solvent is then preheated and distilled to gen-erate renewed solvent at the top of the column. The renewed solvent vapors are condensed to a liquid for storing until ready for reuse. The residual oil in the still bottom is delivered to an accredited waste oil hauler for further processing, it is if found to be nonhazardous.

Bechtel considers recycling "as any activity which reduces hazardous waste volume and/or toxicity with the attendant generation of a valuable substance or an energy stream which is subsequently utilized." Further, Bechtel investigators point out that recycling may take any of the following three forms: (1) direct reuse in the same or another process, (2) reclamation by recovering secondary materials for another use, or (3) regeneration by removing impurities to produce a reasonably pure and reusable sub-stance. They prefer recycling to treatment, but not to source reduction.

Alliance Corporation recommends factors influencing waste reduction techniques.[15] Among these are site-spe-cific factors such as waste volume, waste characteristics, and availability of existing onsite facilities and technical expertise. Small businesses especially fare poorly in waste reduction and tend to land-dispose too much waste. How-

---

[15]Alliance Technologies Corporations *Case Studies of Existing Treatment Applied to Hazardous Waste Banned from Landfill* (Cincinnati, Ohio: EPA PB89-224091, October 1986).

ever, Alliance reports that solvent wastes having a total organic content of greater than 10,000 ppm are the first wastes to be banned from land disposal. Since halogenated-type solvent wastes evolve in smaller volumes and are more expensive to buy (as compared with other solovents), they lend themselves to favorable waste minimization and recycling technologies.

Resource Recovery from Municipal Solid Wastes

Systems involving recovery of resources from solid wastes depend primarily on economics and local situations, such as characteristics of the refuse and land available for landfilling. Most systems utilize sanitary landfilling or incineration as tried-and-true ultimate disposal methods of the residual solids after recovery. At least eight potential overall systems are used:

1. Incineration with recovery of materials from the ash residue
2. Incineration with heat recovery to produce industrial steam
3. Incineration with heat recovery and recovery of materials from ash residue
4. Incineration with heat recovery used to generate electric power for industry or within an existing municipal or industrial utility
5. Pyrolysis, with recovery of oil, char, and inorganic materials from the nonbiodegradable solids
6. Composting that produces humus and inorganic materials from the nonbiodegradable solids
7. Material recovery (paper, aluminum, ferrous metals, and glass) involving separation of mixed refuse into its marketable parts

8. Recovery of organics for use in public utility boiler furnaces as supplemental fuel and ferrous metal recovery

The average composition of municipal refuse in the United States can be assumed to approximate the figures in Table 3.6. The compositions shown in the table vary

Table 3.6 Average Municipal Solid Waste Composition

Materials	Percent of Total Generation[a]						
	1960	1965	1970	1975	1980	1985	1988
Paper and Paperboard	34.1	36.8	36.3	33.6	36.6	38.1	40.0
Glass	7.6	8.4	10.4	10.5	10.0	8.2	7.0
Metals							
Ferrous	11.3	9.8	10.3	9.6	7.8	6.7	6.5
Aluminum	0.5	0.5	0.7	0.9	1.2	1.4	1.4
Other Nonferrous	0.2	0.5	0.6	0.7	0.7	0.6	0.6
Total Metals	12.0	10.7	11.6	11.2	9.7	8.8	8.5
Plastics	0.5	1.4	2.5	3.5	5.2	7.2	8.0
Rubber and Leather	2.3	2.5	2.6	3.0	2.9	2.4	2.5
Textiles	1.9	1.8	1.6	1.7	1.7	1.7	2.1
Wood	3.4	3.4	3.3	3.4	3.3	3.3	3.6
Other	0.1	0.3	0.7	1.3	1.9	2.1	1.7
Total Nonfood Product Wastes	61.8	65.3	69.0	68.3	71.3	71.7	73.5
Other Wastes							
Food Wastes	13.9	12.3	10.5	10.5	8.8	8.2	7.4
Yard Wastes	22.8	20.9	19.0	19.7	18.4	18.6	17.6
Miscellaneous Inorganic Wastes	1.5	1.5	1.5	1.6	1.5	1.5	1.5
Total Other Wastes	38.2	34.7	31.0	31.7	28.7	28.3	26.5
Total MSW Generated —Percent	100.0	100.0	100.0	100.0	100.0	100.0	100.0

[a]Generation before materials recovery or combustion. Details many not add to totals due to rounding.
Source: Franklin Associates, Ltd. Prairie Village, Kansas. Prepared for U.S. EPA, 1990.

regionally and seasonally and will change with economic conditions and scientific and industrial progress. For example, the advent of lightweight aluminum cans has increased that component; the increase in home garbage grinders has decreased that portion. However, there is some indication that there is a decrease in garbage grinders for new homes because of the extra cost of sewage treatment. Hence, garbage may once again increase in the near future. Pressure by conservationists for using returnable bottles has decreased the glass portion.

Some discussion about the recovery of refuse materials, prior to using disposal techniques, is in order at this point.

Paper and Paper Products

It has been approximated that about 50 million tons of paper and paper products have been used by Americans annually during the 1970s. About 80% received one-time use and were then discarded. In the late 1970s paper products constituted 40% to 50% of all solid wastes (an even higher percentage than shown in Table 3.6). Of the 40 million tons of paper waste, 25%, or 10 million tons, were recycled. If recycled paper products were more competitive, about 80% of our needs for paper raw materials could be met with waste paper and hence could relieve the pressure on our forest resources. Paper products are among the easiest to recycle. Separation systems are available, but their use must be encouraged by higher prices for reused paper. However, reuse of paper is not without production problems for example, requirements for uniformity of composition and quality as well as dependability of quantity.

The state of Florida is playing an important role in the recycling of valuable industrial materials from municipal wastes. The Solid Waste Management Law of 1988 provides that "the amount of municipal solid waste that would be disposed of in the absence of municipal solid waste recycling efforts undertaken within the county and the municipalities within its boundaries . . . [must be] reduced by at least 30 percent by the end of 1994."[16] Further, the counties cannot use yard waste, white goods, old tires, or construction and demolition debris to account for more than half the goal.

This state found that in just over a decade, the management of solid wastes "has evolved from open dumping and burning in unlined pits to a comprehensive program of lined landfills, waste to energy plants, and recycling."[17] A significant fraction of Florida municipal solid waste (8.3 pounds per person per day) consists of wastes that present special management problems, including the following seven types of materials: (1) 245,000 tons of waste tires, (2) 51 million gallons of used oil, (3) 3 million refrigerators, stoves, and other "white goods," (4) 782,000 tons of ash from waste-to-energy facilities, (5) 3 to 4 million spend lead acid batteries, (6) 50,000 tons of biohazardous (hospital) waste, and (7) 4 million tons of construction and demolition debris.

The state of Florida indicated that recycling of the seven materials is expected to be accomplished in the following ways:

No.	Material	Method of Reuse
1	Tires	Shredding and using in road surfacing
2	Oil	Burned, exported, lubricants

[16]Department of Environmental Regulation, State of Florida, Tallahassee, Fla. Solid Waste Management, 01 / 90 / 5M.
[17]*Solid Waste Management in Florida, 1990 Annual Report*, Tallahassee, Fla.: Department of Environmental Regulation, State of Florida, March 1991).

No.	Material	Method of Reuse
3	White goods	
4	Ash	Soil cement for road base material, aggregate in road asphalt, and making blocks for artificial reefs
5	Batteries	Recycled and reused by battery manufacturers
6	Biohazardous matter	Combusted for energy
7	Demolition matter	Disposed of in separate area of landfills

From July 1989 to July 1990 the state reported the following amounts of recyclable industrial-type wastes:

1	Paper	29.9% of total
2	Metals	7.1% of total
3	Plastics	7.0% of total
4	Glass	3.9% of total
5	Construction material	18.7% of total
6	Tires	1.3% of total
7	Textiles	2.5% of total
8	*Other	29.6% of total

*Includes food wastes, yard wastes, and miscellaneous wastes.

Recovery of Products after Treatment

PYROLYSIS FOR REUSE

Table 3.7 illustrates what pyrolysis of a ton of typical refuse has been shown to yield. Despite these potential recoveries, recycling and reclamation cannot be expected to handle a major portion of municipal solid wastes for at least a few more years beause of both processing problems and lack of markets for the output. A major impetus can

Table 3.7   Pyrolysis of a Ton of Municipal Refuse

	Urban Refuse	Industrial Refuse of Paper, Rags and Cardboard
Solid Residue	154–424 pounds	618–838 pounds
Tar	0.5–6.0 gallons	
Light Oil	1–4 gallons	1.5–3.0 gallons
Liquor	97–133 gallons	68–75 gallons
$(NH_4)_2 SO_4$	16–32 pounds	12–23 pounds
Gas	7,380–18,058 ft^3	9270–14,065 ft^3

Source: N. L. Nemerow, *Industrial Solid Wastes* (Ballinger Publishing, 1984) 87.

be given to reuse if and when a monetary value is given to the environmental consequences of nonrecovery.

COMPOSTING FOR REUSE

Composting is an admirable solution for solid waste treatment because it converts municipal refuse into a reusable product, a useful soil conditioner. For example, in the 1990 Florida solid waste percentages shown on page 68, "Paper" (29.9%) and "other" (29.6%) represent compostable organic matter. This amounts to a total of 59.5%, or about 4.9 pounds for every person per day. Sorting and salvage precede composting. Sorting removes most inorganics, noncombustibles, bulky materials, and salvageable items. Generally, hand sorting is used to salvage reusable products, but sometimes inertial or magnetic separation is used.

The private market for composting is small and not fully developed. Compost derived from municipal refuse is not ideally suited by itself as a fertilizer since it is relatively low in nitrogen, phosphorous, and potash for use in heavy agriculture. Home gardening, landscaping and park maintenance are major uses for such compost. Recycling before and after composting may not always be feasible, but with the development of better markets and the increase in value of both the products and a clean environment, it is only a question of time before it becomes practical.

In Table 3.8 some typical market prices of various commodities, such as food, grains, fats, textiles, metals, rubber, hides, and fuel oil, are given for August 28, 1991. These prices change hourly on each trading day. For example, aluminum ingots sold for 57 1/2¢ to 58 1/4¢, whereas one year earlier they sold for 82¢. Aluminum prices quoted for September 17, 1979, were 58 1/2¢ to 63¢ per pound for 99%-plus pure 50pound ingots. This provides the reader with two different ways for quoting prices of the same commodity on similar days.

It is interesting to see the change that occurred in the price for newspaper (a recyclable commodity) from $35 to $50 per ton from September 1978 to July 1980 and April 1981. Aluminum also increased to 72¢ per pound in July 1980 and to 86.3¢ in 1994 (Table 3.9). Steel scrap, on the other hand, decreased from $103 to $72 per ton in 1980 and increased once again to $126 per ton in 1994. Fluctuations in market values for recoverable solid wastes make the economics of recovery and reuse difficult to predict. However, in the long run, they represent a challenge similar to that encountered in any industrial production venture.

In considering the economics of any recovery of industrial material from refuse, the specifications of the recovered material are of utmost importance. If the recovered material does not meet the product specifications with a minimum of renovation, the reuse process may not be economically feasible.

FERROUS METALS

Ferrous metals retrieved from refuse come mainly from tin cans. The principal market for tin cans is the copper-

Table 3.8   Cash Prices

(Closing Market Quotations)

Grains and Feeds

	Wed	Tues	Yr. Ago
Barley, top-quality Mpls., bu	n2.05-.25	2.05-.25	2.37 1/2
Bran, wheat middlings, KC ton	60.00	61.0-63.0	64.00
Corn, No. 2 yel. Cent-Ill. bu	bp2.44 1/2	2.47	2.52 1/2
Corn Gluten Feed, Midwest, ton	80.0-96.0	80.0-96.0	75.00
Cottonseed Meal,			
Clksdle, Miss. ton	32 1/2-135.	135.00	180.00
Hominy Feed, Cent-Ill. ton	71.00	71.00	84.00
Meat-Bonemeal, 50% pro. Ill. ton	230.-235.	230.-235.	205.00
Oats, No. 2 milling, Mpls., bu	1.33-.37	1.32-.35	1.17
Sorghum, (Milo) No. 2 Gulf cwt	4.95	4.92	4.70
Soybean Meal,			
Decatur, Illinois ton	184.-188.	183.-187.	171.25
Soybeans, No. 1 yel Cent.-Ill. bu	bp5.70	5.73	6.07
Wheat,			
Spring 14%-pro Mpls. bu	3.14 1/4-24 1/4	3.17 1/4-21	1/4 2.98 1/2
Wheat, No. 2 sft red, St. Lou. bu	bp2.81	2.80	2.89 1/2
Wheat, No. 2 hard KC, bu	3.17 1/2	3.17	2.84
Wheat, No. 1 sft wht, del Port. Ore.	3.60	3.59	3.23

Foods

	Wed	Tues	Yr. Ago
Beef, Carcass, Equiv. Index Value,			
choice 1-3,550-700lbs.	104.65	104.50	114.35
Beef, Carcass, Equiv. Index Value,			
select 1-3,550-700lbs.	101.05	101.70	108.25
Broilers, Dressed "A" NY lb	x.5537	.552	5.6021
Broilers, 12-Cty Comp Wtd Av	.5487	.5487	.5512
Comparable, but not exact.			
Butter, AA, Chgo., lb.	1.03 1/4	1.03 1/4	1.03
Cocoa, Ivory Coast, $metric ton	g1,172	1,206	1,396
Coffee, Brazilian, NY lb.	n.66	.66	.86
Coffee, Colombian, NY lb.	n.87	.86 1/2	1.01
Eggs, Lge white, Chgo doz.	.67-.73	.67-.73	.73
Flour, hard winter KC cwt	8.45	8.40	7.65
Hams, 17-20 lbs, Mid-US lb fob	.76	.76	.93
Hogs, Iowa-S.Minn. avg. cwt	48.25	49.25	53.00
Hogs, Omaha avg cwt	48.00	48.35	52.50
Pork Bellies, 12-14 lbs Mid-US lb	.42	.41-.42	z
Pork Loins, 14-18 lbs. Mid-US lb	1.01-.101.	04-.09 1/2	118.25
Steers, Tex.-Okla. ch avg cwt	68.25	69.25	77.00

Table 3.8 (continued)

	Foods		
Steers, Feeder, Okl Cty, av cwt	93.75	95.13	99.50
Sugar, cane, raw, world, lb, fob	.0940	.0953	.1103
	Fats and Oils		
Coconut Oil, crd, N. Orleans lb.	xxn.20 1/2	.21 1/2	.14 3/4
Corn Oil, crd wet mill, Chgo. lb.	.29-.29 1/4	.29 3/4	.28 3/4
Corn Oil, crd dry mill, Chgo. lb.	n.29 3/4	.29 3/4	.28 1/2
Grease, choice white, Chgo lb.	.11 3/4	.11 3/4	.11
Lard, Chgo lb.	n.13 1/2	.13 1/2	.11 1/2
Palm Oil, ref. bl. deod. N.Orl. lb.	n.18	.18 1/4	.15 1/4
Soybean Oil, crd, Decatur, lb.	.1985	.1995	.2465
Tallow, bleachable, Chgo lb.	.13 3/4	.13 3/4	.12
Tallow, edible, Chgo lb.	n.14 1/2	.14 1/2	.13
	Fibers and Textiles		
Burlap, 10 oz 40-in NY yd	n.2725	.2725	.2875
Cotton 1 1/16 str lw-md Mphs lb	.7003	.6892	.7576
Wool, 64s, Staple, Terr. del. lb.	2.10	2.10	2.35
	Metals		
Aluminum			
ingot lb. del. Midwest	q57 1/2-58 1/4	.57 1/2-58 1/4	.82
Copper			
cathodes lb.	p1.07-.09	1.06 1/8-.08	1.33
Copper Scrap, No 2 wire NY lb	k.83	.83	1.02
Lead, lb.	p.33	.33	.51
Mercury 76 lb. flask NY	q95.-105.	95.-105.	240.00
Steel Scrap 1 hvy mlt Chgo ton	94.00	94.00	118.00
Tin composite lb.	q3.6105	3.6199	3.8558
Zinc Special High grade lb	q.48000	.48000	87 1/4
	Miscellaneous		
Rubber, smoked sheets, NY lb.	n.44 1/4	.44 1/4	.49
Hides, hvy native steers lb., fob	.75	.75	.90 1/2
	Precious Metals		
Gold, troy oz			
Engelhard indust bullion	355.75	355.05	385.60
Engelhard fabric prods	373.54	372.80	404.88
Handy & Harman base price	354.50	353.80	384.30
London fixing AM 353.30 PM	354.50	353.80	384.30

Table 3.8 (continued)

Precious Metals

Krugerrand, whol	a354.50	355.00	387.00
Maple Leaf, troy oz.	a362.50	363.00	399.00
American Eagle, troy oz.	a366.50	367.00	399.00
Platinum, (Free Mkt.)	341.00	333.00	483.00
Platinum, indust (Engelhard)	341.00	336.00	485.00
Platinum, fabric prd (Engelhard)	441.00	436.00	585.00
Palladium, indust (Engelhard)	83.00	82.00	113.25
Palladium, fabric prd (Engelhard)	98.00	97.00	128.25
Silver, troy ounce			
Engelhard indust bullion	3.955	3.950	4.845
Engelhard fabric prods	4.232	4.227	5.184
Handy & Harman base price	3.920	3.930	4.810
London Fixing (in pounds)			
Spot (U.S. equiv. $3.9500)	2.3570	2.3555	2.4925
3 months	2.4195	2.4185	2.5845
6 months	2.4700	2.4770	2.6740
1 year	2.6050	2.6035	2.8530
Coins, whol $1,000 face val	a2,897	2,889	3,485

a-Asked. b-Bid. bp-Country elevator bids to producers. c-Corrected. d-Dealer market. e-Estimated. f-Dow Jones International Petroleum Report. g-Main crop, ex-dock, warehouses, Eastern Seaboard, north of Hatteras. i.-f.o.b. warehouse. k-Dealer selling prices in lots of 40,000 pounds or more, f.o.b. buyer's works. n-Nominal. p-Producer price. q-Metals Week. r-Rail bids. s-Thread count 78x54. x-Less than truckloads. z-Not quoted. xx-f.o.b. tankcars

Source: The New York Times, Business Section, 28 August 1991.

Table 3.9 Cash Prices

	Wed.	Tue.

Foods

Butter AA Chi. lb.	.75 3/4	75 3/4
Broilers dressed lb.	.4831	.4817
Eggs large white NY Doz.	.67	.67
Flour Minn. Std Spring Patent cwt	15.20	15.25
Coffee parana ex-dock NY per lb.	1.70	1.65
Coffee medlin ex-dock NY per lb	1.85 1/2	1.80 1/2
Cocoa beans Ivory Coast $ metric ton	1605	1630
Cocoa butter African styl $ met ton	3656	3734
Sugar No. 11 cents per lb	13.66	13.83
Hogs Omaha 1-2 200-250 lb avg cwt	27.50	29.00

Table 3.9 (continued)

	Wed.	Tue.
Foods		
Feeder cattle 500-600 lb Okl av cwt	82.00	82.00
Pork bellies 12-14 lb Midwest av cwt	.31	.31
Grains		
Corn No. 2 yellow Chi processor bid	2.16 3/4	2.15 3/4
Soybeans No. 1 yellow	5.55 1/2	5.56
Soybean Meal Cen Ill 48pct protein-ton	159.50	160.50
Wheat No. 2 Chi soft	3.78 3/4	3.83 1/4
Wheat N. 1 dk 14pc-pro Mpls.	4.31 1/4	4.25 1/2
Wheat No. 2 hard KC.	4.12	4.19
Oats No. 2 heavy Mpls	n.q.	n.q.
Fats and Oils		
Coconut oil N. Orleans lb.	.34 1/2	.34
Corn oil crude dry mill Chi. lb.	.27 3/4	.27 3/4
Soybean oil crude Decatur lb.	.30 3/8	.30 1/4
Metals		
Aluminum cents per lb LME	86.3	84.3
Antimony bulk 99.5 pct NY per lb.	2.25	2.25
Copper Cathode full plate	133.00	129.00
Gold Handy & Harman	386.25	386.70
Silver Handy & Harman	5.240	5.215
Lead per lb.	.44	.44
Pig Iron fob fdry buff gross ton	213.00	213.00
Platinum per troy oz. NY (contract)	407.00	407.00
Platinum Merc spot per troy oz.	416.40	417.60
Mercury per flask of 76 lbs.	215.00	215.00
Steel scrap No. 1 heavy gross ton	126.00	126.00
Tin Metals Week composite lb.	4.1801	4.1303
Zinc (HG) delivered lb.	.5905	.5796
Textiles and Fibers		
Cotton 1-1-16 in. strict low middling	72.50	71.14
Wool fine staple terr Boston lb.	2.30	2.30
Miscellaneous		
Rubber No. 1 NY smoked sheets lb.	.71 3/4	r.71 1/2
Hides heavy native steer lb.	.94	.94

Table 3.9 (continued)

	Wed.	Tue.
Petroleum—Refined Products		
Fuel oil No. 2 NY hbr bg gl fob	.4755	.4840
Gasoline unl prem RVP NY hbr bg gl fob	.6300	.5780
Gasoline unl RVP NY hbr bg gl fob	.5660	.5495
Prices provided by Oil Buyer's Guide		
x- prices are for RVP grade of gasoline		
Petroleum—Crude Grades		
Saudi Arabian light $ per bbl fob	15.45	15.60
North Sea Brent $ per bbl fob	16.80	16.45
West Texas Intermed $ per bbl fob	17.35	17.60
Alask No. Slope del. US Gulf Coast	16.20	15.60
RawProducts		
Natural Gas Henry Hub, $ per mmbtu	n.a.	n.a.
b—bid a-asked.		
n—Nominal		
r—revised.		
n.q.—not quoted.		
n.a.—not available.		

Source: *The New York Times*, Business Section, 17 November 1994.

smelting industry (where tin contamination is not a hindrance). The costs associated with separating, recovering, and processing scrap may, in some cases, surpass the value of the scrap. Once again economics prevail, and the cost of environmental degradation by tin cans is not included in the price of recovery and reuse.

ALUMINUM

The use of aluminum cans has increased, and so has the value of recovered and reused cans. The production of raw aluminum from bauxite is expensive and uses large amounts of energy, as shown later in Table 3.10. A method of separating aluminum from other nonferrous metals must be developed. Collection locations have been established for the return of aluminum cans, with industry

Table 3.10 Energy Saved by Recycling

	Energy Required (kwh/ton)	
	Original Production	Recycling
Steel	4.3	1.7
Copper	13.5	1.7
Aluminum	51.4	2.0
Paper	5.0	1.5

Source: N. L. Nemerow, *Industrial Solid Wastes* (New York: Ballinger, 1984), 29.

(Reynolds Aluminum Co.) assisting. In Miami, Florida, Reynolds offered $.27 per pound for discarded aluminum cans as long as 14 years ago (see *The Miami Herald*, Neighbors Section, 22 January 1981).

OTHER NONFERROUS METALS AND ASH

Lead, copper, zinc, and tin all have high salvage values. However, they are generally wasted by industries and municipalities in small quantities and, as impure alloys, their recovery is difficult.

One method being researched and piloted for separating these valuable metals from ferrous metals and nonmetallic solids is that of cryogenics. At low temperatures some materials become brittle, while other materials remain malleable. Crushing the frozen material to shatter the embrittled matter and subsequently subjecting it to magnets allows screening or gravity separation to be used to recover specific components. Dry ice (solid CO_2) can produce temperatures low enough to effect acceptable separation of insulated wire and mixed nonferrous refuse.

Fly ash and bottom ash from power plants are now being reused. For example, recovered fly ash from stack gases can be used as a cement additive in the manufacture of concrete blocks.

GLASS

Glass has a low salvage value, and industrial processors have been resistant to reusing it unless it is of one color and grade and is free from impurities such as metals. Economic limitations (low glass value) seem to control its recovery potential. Sand is not very expensive as a raw material in glass manufacture, but fuel and energy to fuse it into the product may be sufficiently important and costly to justify the recovery of glass.

One use of glass recovered from incinerator residue is in the manufacture of glass wool. This material is used for attic and ceiling insulation in houses and for blanket insulation in building walls. It can also be used as a flux for common clays. When added to clay before it is cured, it can lower the maturing temperature significantly to save a considerable amount of kiln fuel, usually natural gas. Glass has also been used successfully as an aggregate additive for highways. It has been reported to increase the life and structural characteristics of roads, and because glass sparkles, it can aid the driver's vision.

RUBBER

Rubber, usually in the form of old tires, is easily separated from refuse. However, because of the great variety in type and grade of tires, no large-scale recycling is taking place. The increase in the use of a variety of steel-belted radial tires has made the recovery of rubber all the more difficult. Tires do burn with very high heat, however, and after shredding, can be used as fuel for boilers to recover heat or power.

PLASTICS

An increasingly greater percentage of refuse is made up of plastics, because of their utility in modern, affluent societies. Most are non biodegradable and, hence, landfilling and composting of plastics are impractical. Recovery of almost all plastics can be accomplished by electrodynamic separation techniques. Some plastics, when incinerated, yield HCl and contaminate the air surrounding such operations.

AUTOMOBILES

The salvage of old automobiles is now profitable. The major problem in recovering valuable components from autos (steel, iron, rubber, etc.) is the difficulty in efficient separation of the various components. Crushers have been designed and constructed for the recovery of scrap metal in its dense form, and magnetic separators aid in separating ferrous from nonferrous metals. Higher prices of scrap iron for the steel industry have made this salvage economically feasible. Reuse potential depends on the health of the steel industry. However, the recovery of other auto components has not yet been practical or economical.

Recovery is practiced by both auto dismantlers and scrap processors. The former are retailers who buy vehicles to obtain usable parts again for resale. The dismantler must be concerned with year, make, model, and condition of the autos purchased. After the salable parts have been removed, the remainder is sold by weight to the scrap processor, who is purchasing unprepared scrap for processing and final recycling.

A $2.5 million plant has been designed, using a 2,400 horse power hammermill constructed to grind up 6,000 cars a day into 90-pound chunks of steel and smaller quantities of aluminum, brass, and copper (See *The Miami Herald*, 10, November 1986, page 6). This scrap metal company pays scavengers of junk cars about $2.50 per 100 pounds. The company expects the average 3,000-pound old car, minus the fuel tank, to yield approximately 2,250 pounds of ferrous metals, 90 pounds of nonferrous metals, and 660 pounds of nonmetallic waste. The ferrous scrap is extracted from the shredded debris by giant electromagnets, and nonferrous metals are segregated in a series of flotation tanks. Copper, aluminum, and zinc are sold separately to metal brokers. The company also expects to install equipment designed to reclaim platinum from catalytic convertors. Table 3.11 lists some typical material specifications for those items usually recovered.

Many consultants and urban agencies conclude that incineration and composting are losing favor as treatment methods because of technical, economical, and environmental considerations. These include (1) the increase in capital and operating costs of such treatments, at least partially resulting from mandated environmental controls and (2) the decrease in potential value of the constituents present in solid wastes, such as Btu value for incineration and organic matter for composting. This trend may be caused by an increased interest in conservation of resources such as paper and plastics. Thus, our objectives are turning to (1) developing new technology aimed at recovering raw materials for the energy-producing portion of urban solid waste, (2) reducing the quantity of contaminants to both water and air resulting from other methods of disposals, and (3) obtaining methods for using solid waste constituents for resale, including that type which is amenable to energy production.

Recycling signifies a return to the consumption cycle of materials (partially or completely renovated and finished) recovered from solid wastes. This can be accomplished by either direct or indirect means. The former implies reuse of materials without altering their physical, chemical, or biological character. For example, the reuse of broken glass by the glass industry, or the reuse of paper to make pulp by the paper industry, represent direct recycling. Indirect recycling involves transforming the recovered

Table 3.11 Typical Materials Specifications That Affect Selection and Design of Processing Operations

Reuse Category and Materials Component	Typical Specification Items
	Raw Material
Paper and Cardboard	Source; grade, no magazines; no adhesives; quantity, storage, and delivery point
Rubber	Recapping standards; specifications for other uses not well defined
Plastics	Type (for example, ABS, PVC); degree of cleanliness
Textiles	Type of material; degree of cleanliness
Glass	Amount of cullet material; color, no labels or metal; degree of cleanliness; freedom from metallic contamination; quantity, storage, and delivery point
Ferrous Metals	Source (domestic, industrial, etc.); density; degree of cleanliness; degree of contamination with tin, aluminum, and lead; quantity, shipment means, and delivery point
Aluminum	Particle size; degree of cleanliness; density; quantity, shipment means, and delivery point
Nonferrous Metals	Vary with local needs and markets
	Fuel Source
Combustible Organics	Composition, Btu content; moisture content; storage age limits; firm quantities; sale and distribution of energy or by-products
Wastepaper	Vary with local needs and markets
	Land Reclamation
Organics	Local and state regulations; method of application; control of methane gas migration; leachate control; final land-use designation
Inorganics	Local and state regulations; final land-use designation

material so that it can be reused for a purpose different from its original use. For example, after crushing, glass can be reused as filling material for construction items such as roads or building blocks and organic matter can be used as a fertilizer after composting. Recycling materials is also known to save energy, as shown in Table 3.10. In reusing aluminum, a major energy consumer, about 97% energy is saved during production.

Because recovery processes vary in efficiencies, not all components can be recovered completely. In Table 3.12, some indication of these efficiencies is presented. These materials must pass through several types of processing in order to be removed as effectively as possible. Some of these processes are discussed in the following section.

PROCESS FLOWSHEETS

Dry-process flowsheets are presented in Figures 3.5 and 3.6. Dry processing is less costly than wet processing,

Table 3.12 Estimated Recoverable Quantities for Various Components in Solid Wastes Using Mechanical Equipment

Fraction or Component	Recoverable Portion of Original Components, Percent		Comments
	Range	Typical	
Light fraction	80–95	90[a]	Recoverable portion will vary with the composition of the solid wastes and the characteristics of the wastes after shredding.
Heavy fraction	90–98	96[b]	
Ferrous metal	65–95	85	Varying amounts of light and heavy fraction material will also be removed with these components depending on the specific process and equipment used.
Glass	50–90	80	
Aluminum	55–90	70	

[a]Varying amounts of the light fraction will be retained with the heavy fraction.
[b]Varying amounts of the heavy fraction will be carried over with the light fraction.

because a hydropulper is used in the wet process. Standard equipment used in mineral-processing industries can be adopted for use in the dry process. In both flowsheets, air classification follows primary shredding, and cyclone separators remove the contaminated air from the light fraction. From a review of Figures 3.5 and 3.6, it is evident that many variable flowsheets can be prepared. Manual separations of specific waste components are also commonly used. Important factors that must be considered in the design and layout of such systems include (1) process performance efficiency, (2) reliability and flexibility, (3) ease and economy of operation, (4) aesthetics, and (5) environmental controls. The degree of separation achieved for the various components determines the efficiency of the system. The quantity of material to be handled must be determined as a first step in the design of a processing facility. In situations where the processed wastes are to serve as a source of fuel, the design quantities usually depend on the amount of continuous power that must be developed. Units are sized according to the loading rates, which are determined on the basis of the characteristics of the solid waste and the separation process to be used. Estimation of the quantities of materials that can be recovered and of the appropriate design loading rates is an important part of any recovery system. Data and other information that can be used to estimate the required quantities are presented in Table 3.12. The components that normally make up the light and heavy fractions after shredding and air classification are identified in Table 3.13.

Belt and chain belt conveyors of unprocessed solid wastes have proved especially troublesome. Conveyors can easily and frequently be damaged by heavy, bulky, and cumbersome solid wastes dropped onto them. Wire and rope in the wastes become snagged on the equipment,

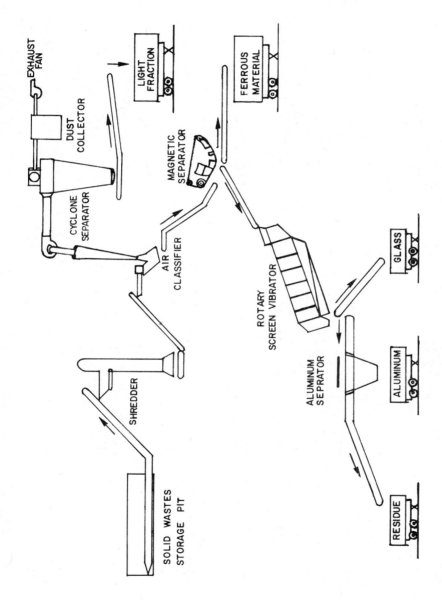

Figure 3.5. Pictorial Flowsheet for Materials Recovery Systems

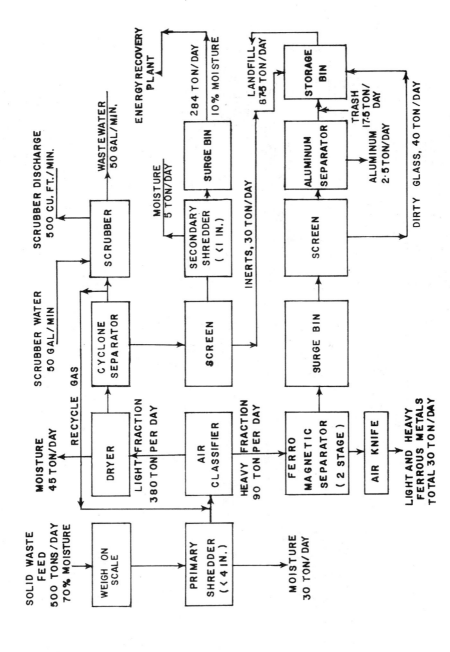

Figure 3.6. Flowsheet for Materials Recovery System

84

Table 3.13 Light and Heavy Fractions of Solid Waste Components After Shredding

Component	Percent by Weight[a]	Fraction by Weight, Percent		Comment
		Light	Heavy	
Food wastes	15	15	—	
Paper	40	40	—	Components assumed to make up
Cardboard	4	4	—	the light fraction after shredding.
Plastics	3	3	—	After air classification the light
Textiles	2	2	—	fraction will contain from 2% to 8%
Rubber	0.5	0.5	—	of the components from the heavy
Leather	0.5	0.5	—	fraction by weight.
Garden trimmings	12	12	—	
Wood	2	2	—	
Glass	8	—	8	Components assumed to make up
Tin cans	6	—	6	the heavy fraction after shredding.
Nonferrous metals	1	—	1	After air classification the heavy
Ferrous metals	2	—	2	fraction will contain from 5 to 20%
Dirt, ashes, brick, and so on	4	—	4	of the light fraction components by weight.
Total	100	79	21	

[a]Moisture loss during shredding not considered.

causing waste spillage and overflows, comparable to the way a clogged or obstructed sewer backs up flow into basements and streets. The binding and wedging of conveyor systems has also been a problem. Because of the abrasive nature of many components such as metal parts —some broken and jagged—found in solid wastes, processing equipment wears out rapidly.

Independent operation of duplicate units is now recommended, especially where continuous power is produced from the heat generated. Equipment should be selected for easy repair, with standard parts and components. Sprockets, gears, pins, and belts are especially troublesome. Oils, gases, organic compounds, and heat can be produced by the chemical conversion of solid wastes. Incineration and pyrolytic processes are the few full-scale installations that have been producing these by-products. With pyrolysis, most of the full-scale experience—exclusive of pilot plants —is in the petroleum and wood-processing industries.

Because recovery processes vary in efficiencies, not all components can be recovered completely. Table 3.12 gives some indication of these efficiencies. The materials shown in Table 3.12 must pass through several types of processing in order to be removed as effectively as possible. We show some of these processes in the flowsheets illustrated in Figures 3.5 and 3.6. Some of the chemical processes that can be used for the conversion of solid wastes are listed in Table 3.14.

Enadisma (a firm in Spain engaged in recycling) proposes a resource recovery process of pneumatic separation accompanied by the basic operation of grinding and magnetic and complementary mechanical separation procedures. A schematic drawing of the process is shown in Figure 3.7.

The recovery and reuse of solid wastes is not always without its problems and its own wastes. Many times, recovery techniques result in further pollution of the environment, and often by insidious discharge of hazardous wastes. Such a situation was illustrated by the Sapp Battery Plant of Alford, Florida (as reported in The *Miami Herald*, 26 July 1981), which recovered spent auto batteries. The recovery plant removed the tops of as many as 50,000 old batteries, per week, also removed the metal plates in each battery, and dumped out deposits of bottom sludge. The heavy plates and sludge contained lead that Sapp sold to smelting companies for about $1 million per month. The guts of the batteries awash in the water from spraying pits, were heaped outside the plant. The water drained into a pond on Sapp's property, then found its way into a culvert that ran under a dirt road, and trickled into

Table 3.14 Chemical Processes Used for Conversion of Solid Wastes

Process	Conversion Product	Preprocessing Required	Comment
Incineration with heat recovery	Energy in the form of steam	None	Markets for steam must be available; proved in numerous full-scale applications; air-quality regulations may prohibit use.
Supplementary fuel firing	Energy in the form of steam	Shredding, air separation, and magnetic separation	If least capital investment desired, existing boiler must be capable of modification; air-quality regulations may prohibit use.
Fluidized bed incineration	Energy in the form of steam	Shredding, air separation, and magnetic separation	Fluidized bed incinerator can also be used for industrial sludges.
Pyrolysis	Energy in the form of gas or oil	Shredding, magnetic separation	Technology proved only in pilot applications; even though pollution is minimized, air-quality regulations may prohibit use.
Hydrolysis	Glucose, furfural	Shredding, air separation	Technology on laboratory scale only.
Chemical conversion	Oil, gas, cellulose acetate	Shredding, air separation	Technology on laboratory scale only.

a swamp. Gradually, the surrounding trees began to take on a steely look, turning from green to brown and, finally, to cold, gray stumps. Sapp's 40 acres had become a "sponge" soaked with sulfuric acid, lead, and heavy metal residues. When it rained, the contaminants seeped out into the creek and flowed 50 miles downstream into a river and a series of lakes. The contaminants killed trees and fish in the creek, and unusually high levels of lead

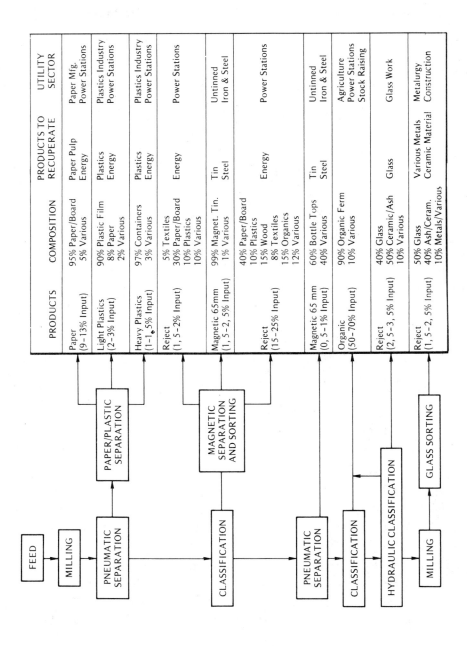

PRODUCTS	COMPOSITION	PRODUCTS TO RECUPERATE	UTILITY SECTOR
Paper (9–13% Input)	95% Paper/Board 5% Various	Paper Pulp Energy	Paper Mfg. Power Stations
Light Plastics (2–3% Input)	90% Plastic Film 8% Paper 2% Various	Plastics Energy	Plastics Industry Power Stations
Heavy Plastics (1–1,5% Input)	97% Containers 3% Various	Plastics Energy	Plastics Industry Power Stations
Reject (1,5–2% Input)	5% Textiles 30% Paper/Board 10% Plastics 10% Various	Energy	Power Stations
Magnetic 65mm (1,5–2,5% Input)	99% Magnet. Tin. 1% Various	Tin Steel	Untinned Iron & Steel
Reject (15–25% Input)	40% Paper/Board 10% Plastics 15% Wood 8% Textiles 15% Organics 12% Various	Energy	Power Stations
Magnetic 65 mm (0,5–1% Input)	60% Bottle Tops 40% Various	Tin Steel	Untinned Iron & Steel
Organic (50–70% Input)	90% Organic Ferm 10% Various		Agriculture Power Stations Stock Raising
Reject (2,5–3,5% Input)	40% Glass 50% Ceramic/Ash 10% Various	Glass	Glass Work
Reject (1,5–2,5% Input)	50% Glass 40% Ash/Ceram. 10% Metals/Various	Various Metals Ceramic Material	Metalurgy Construction

Figure 3.7. Schematic of Process for Treatment of Uniform Solid Waste (Enadimsa) Composition and Utility Alternatives

and heavy metals have been found in the tissues of clams in the lakes. The Florida Department of Environmental Regulation assessed a $11,159,940 judgment on Sapp-$4.5 million in punitive damages and $6.4 million in actual damages to the environment. Whether the lead can ever be reclaimed and the area returned to an uncontaminated state is unknown. Even plants recovering and reusing contaminants must be aware of their subsequent pollution potential.

Some waste materials are recycled more as a necessity rather than for economic gain. For example, the disposal of polychlorinated biphenols (PCBs) found in defunct fluorescent light ballasts can be expensive if not handled properly. Dong and Mc Cagg have found about one ounce of virtually pure PCB in each ballast. They give seven options for disposal of the used PCB (of these, only incineration—with and without recyclying—is acceptable): (1) leaving the disconnected PCB ballast in the ceiling, (2) sanitary landfill, (3) municipal incinerator or resource-recovery facility, (4) chemical waste landfill, (5) chopping open the ballast (none of these five methods is recommended by Dong and Mc Cagg); (6) whole ballast incineration in a PCB incinerator, and (7) capacitor removal/incineration and recycling. The seventh method is highly recommended by these authors. They do this by removing the 4-ounce capacitor and asphalt potting material from the 3.5-pound ballast, incinerating only the PCB-contaminated material and recycling the remaining uncontaminating metals (copper, steel, and aluminum). They claim that more than 80% of the ballast weight in recoverable. This system is 30% to 50% less costly than whole-ballast incineration, generates less air pollution and ash, and uses less fuel, while reclaiming valuable metals.[18]

---

[18]M. Dong, and B. McCagg, "Ful Circle's Practical Guide to PCB Ballast Disposal," *Bulletin of Ful Circle Recyclers, Inc.* (August 1993), Bronx, N.Y.

D. Banker reports on recycling incinerator ash, another recently designated hazardous waste. Wheelabrator Environmental Systems plans to mix incinerator ash with cement and sell it to contractors, who will use the crushed mix in place of limestone and combine it with asphalt to create blacktop. Despite ash's classification as a hazardous waste, no adverse environmental impact has been found to result from its use as an additive for road surfacing. Fly ash, which is considerably more toxic than bottom ash, accounts for only 10% to 20% of incinerator ash, and the remainder is less toxic bottom ash. Wheelabrator will sell the ash to road builders for about half the current price ($2.00 to $3.00 per ton) of road building mix.[19]

---

[19]D. Banker, "Ash from Trash in Broward Could Be Used to Pave Roads," *The Miami Herald*, 5 June 1994, 1B.

CHAPTER 4

WASTE MINIMIZATION BY RECOVERY AND EXTERNAL SALE OF PRODUCTS

DIRECT SALE TO OTHERS

Recovery of waste products of an industrial production process is a feasible method for avoiding both pollution and excessive waste treatment costs. This procedure is not used more often or more generally because it usually means more work for an industry outside the normal flow of production for profit.

However, there are many examples of such waste minimization, including (1) the sale of feathers recovered at chicken processing plants to pillow manufacturers for stuffing purposes, (2) the sale of grease by abattoirs to rendering plants for manufacturing soaps, and (3) the sale of culls from cannery waste solids for use by farmers in feeding animals. Note that each of these waste products is reusable directly by another industry without any further treatment or preparation.

Other classic examples include the sale of waste pulp-mill digestion liquors for use by food manufacturers in making vanillin, and the sale of steel-mill slag waste to cement manufacturers to make cement products, railway ballast, an aggregate for concrete, or even a filter medium for sewage disposal plants or an agricultural fertilizer.

For other examples of this method of waste minimization, refer to my 1984 book, *Industrial Solid Wastes* (Ballinger). The purpose here is to show the utility of such procedures and to discuss their major ramifications.

In Collier County, Florida, reclamation has progressed even to the point of recovering and recycling materials from old municipal landfills. The system of landfill mining transfers waste from old landfills with a front-end loader onto equipment that separates the soil-organic material from larger item such as tires and white goods (refrigerators, washing machines, etc.). Ferrous metal is extracted from the solids flow for recycling. Residual solids are sent to recyclers or, if not acceptable for recycling, returned to a new lined landfill. Investigators claim that there are four advantages to this mining process: (1) recovery of soil for reuse as soil cover, (2) recovery of recyclable materials, (3) better control of leachates, and (4) reuse of the land.

Early data from this landfill-mining reuse project indicate that 75% of the mass in completed landfills can be recovered as backfill and used as construction or cover material for other landfill cells. The researchers suggest that by using aerobic degradation, landfill mining—operated as a system—could extend a landfill's life by three to five years. They believe that old landfills can be "retrofitted to include this type of evolving technology." Their opinion is that when landfills leak, it would be better to remove the material, recycle what is possible, and landfill the remainder on a lined site.[1]

Recycling can be done both on-site and off-site. The following advantages and disadvantages of both types of recycling are adapted from Bechtel.[2] These provide the reader with valuable clues to their utility.

Advantages of On-Site Recycling	Disadvantages of On-Site Recycling
1. Lessens overall waste lost	1. Higher capital costs for recycling equipment
2. Better control of the recycling process and quality of recycled material	2. Higher operating and maintenance costs for recycling equipment
3. Lowered liability since waste material does not require off-site transportation	3. Higher utilities cost for recycling equipment

[1] R. G. Haines, "Researcher: There's Wealth in Those (Landfill) Hills". *Business Weekly* (April 1991), 21.
[2] *Waste Minimization Study for the Lawrence Levermore National Laboratory: Solvent Wastes* (July 31, 1987), Bechtel National, Inc., P.O. Box 3965, San Francisco, CA 94119.

Advantages of On-Site Recycling	Disadvantages of On-Site Recycling
4. Less paperwork than required to ship off-site	4. Increased need for operator training
5. Possible lower unit cost for recycled waste,	5. Increased risk to recycling equipment
6. eliminating the need for federal or state permits	

Advantages of Off-Site Recycling	Disadvantages of Off-Site Recycling
1. avoids capital costs of recycling equipment	1. Higher transportation costs for recycler
2. avoids operating and maintenance	2. Increased liability of industry for recycler-generated problems
3. avoids utilities costs for recycling equipment	3. Cradle-to-grave liability responsibility for industrial generator
4. avoids need for operating training	4. Increased paperwork required of industrial generator before the waste leaves site for recycling
5. avoids risk of operating recycling equipment	

INDIRECT SALE TO OTHERS THROUGH REGIONAL EXCHANGES

The concept of regional exchanges or markets for transferring hazardous materials from one supplier to another user is a rather new and intriguing potential solution to a dilemma facing society. A regional center enhances the exchange of materials, rather than allowing their alternative disposal into the environment. This process requires a great deal of communication with, confidence in, and consideration of all parties concerned. Advertisement of products, materials, and chemicals available for the exchange must be extensive but, at the same time, discreet. Sometimes buyers and sellers must remain anonymous. The exchange must have competent and trustworthy management, somewhat like a stock market exchange but without the usual product exposure to the general public. Certainly, the objective of reuse without discharge is an admirable one. However, the procedure needs refinement, practice, and much experience.

There is a similarity between the objectives of these exchanges and those of zero pollution attainment described in Chapter 15 of Industrial Hazardous Waste Treatment.[3] The zero pollution goal, however, is attained by all manufacturing of related products taking place in the same industrial complex.

E. LeRoy reports that exchanges have been set up in France through local initiative. Their main purpose is to create and develop new outlets for using waste and saving raw materials.[4]

The National Conference on Waste Exchange defines a waste exchange as "an operation that engages in transformation of either information concerning waste materials or the waste materials themselves."[5] Waste exchanges were first organized in Europe where depletion of readily available natural resources and limited land disposal forced manufacturers to find alternative sources of raw materials. All of the foreign exchanges are information clearinghouses, and most are operated by trade organizations, primarily in the chemical industry. Material exchanges have been the major innovation in waste avoidance concepts that developed in the United States in the mid 1970s.

---

[3] N. L. Nemerow and A. Dasgupta, *Industrial Hazardous Waster Treatment* (New York: Van Nostrand, Reinhold, 1991).

[4] E. LeRoy, "Processing Hazardous Waste in France," *Industry and Environment Hazardous Waste Management*, UNEP special issue, No. 4 (1983): 46.

[5] *Proceedings of the National Conference on Waste Exchange*, Florida State University, Tallahassee, Fla. March 8-9, 1983, p. 82.

Surplus materials and equipment, scrap metals, and discontinued products, as well as traditional wastes, may enter the waste exchange cycle. Industries are using the term "investment recovery" to describe the entire process of recycling and reuse within plants. Some firms have already investigated or instituted process modifications that have proven successful, which include (1) substitution of reclaimed acid for typical new electroplating acids, (2) source separation and segregation of various materials, (3) concentration and volume reduction, (4) altering raw materials specifications to allow substitution of lower-quality inputs, (5) using intermediate reactions designed to modify waste stream components, (6) tightening process control to take advantage of by-product (not waste) streams, and (7) educating plant management and workers on the benefits of resource reuse.

A deterrent to successful waste exchange programs is the possibility of exposing private information to persons outside a particular industrial plant. The program must maintain an agreement of "confidentiality" to protect the proprietary interests of generators and to limit the direct identification of specific firms that are generating a particular material. Industry identifies "potential liability" as the primary reason for nonparticipation in waste transfer agreements.

Four requirements have been identified for industrial resource reuse and waste transfer.[6]

1. Participation in a waste exchange must be uncomplicated and cost-effective. The exchange itself must be reputable and reliable.
2. Alternatives to conventional treatment and disposal methods that are presented by waste exchange must be ethical and cost-effective

---

[6] Ibid.

3. The exchange should have as wide an audience of potential users as possible and should have extensive contacts in the waste management field in order to be aware of all waste management options.
4. The generator must know where and in what form his waste is being reused or disposed of.

Many areas of cooperation between exchanges are possible, including the following: (1) common database shared on the regional or national level, (2) trading of listings between exchanges for catalog distribution, (3) network of regional contacts for information referral, and (4) licensing of exchanges to ensure maintenance of quality of service. However, industrial managers hesitate to participate in cooperative exchanges because of their potential liability for mismanagement of waste. Unfavorable economics, when effected by high transportation costs for hazardous wastes, may be another deterrent to involvement in cooperative exchanges.

It was reported by the Aregonne National Laboratory[7] that the amount of energy saved by one waste exchange over two and a half years was 10 x 10^9 Btu's. The investigators calculated that a savings of 10^{12} Btu's per year (the equivalent of 100,000 barrels of oil at 10^7 Btu's per barrel) would result if 50 exchanges as large or as effective as the one studied existed.

As with other liquid, gaseous, and solid wastes, there are generally no tax incentives, credits, or advantages for industries that recycle hazardous wastes. On the other hand, there are taxes levied for producing such pollutants. Exceptions are the states of New Jersey and California, where tax advantages are given (or are being considered) for recycling waste.

---

[7] Ibid.

An example of regional exchange of waste information is the Southern Waste Information Exchange (SWIX), which serves the southern region of the United States. It publishes and distributes to interested firms listings of materials available, materials wanted, services available, and services wanted. Examples of listing forms are shown in Figures 4.1 and 4.2. Materials include acids and alkalies, inorganic chemicals, solvents, oils and paints, paper, wood, plastic, rubber, glass, and miscellaneous items. Services include recycling equipment and supplies, engineering consulting, legal and health-related consulting, tank cleaning and lining, collection and transportation, storage, treatment and disposal, and miscellaneous functions.[8]

In Florida, the state provides the Center for Solid and Hazardous Waste Management, which operates an electronic bulletin board called the Florida Recycling Marketing System (FRMS).[9] The system is nationwide, intended to promote communication and information exchange among the various parties involved in recycling. A primary objective is to bring together sellers and buyers of used materials. It also promotes exchange of information on recycling programs so as to establish a network that can be used to solve problems. As of mid-1991, the bulletin board was used primarily for information on products of paper, plastic, glass, and rubber and will expand to include other industrial materials. All this information is available by using a personal computer with a modem and dialing an 800 number. Once accessed, the system provides the user with simple instructions on how to register and use the various parts of the bulletin board.

---

[8] *The Southern Waste Information Exchange Catalogue (SWIX)* 11, No. 1, (October 1982). P.O. Box 6487, Tallahassee, FL 32313.
[9] Florida Center for Solid and Hazardous Waste Management, Letter information release, August 14, 1991. Gainesville, FL 32608-3848.

Waste Management Services
Listing Form

A separate form is required for each service listed.
Limit of two listings.

1. Company Name: _____

2. Mailing Address: _____

                    _____

                    _____

3. Company Contact:_____   4. Title:_____

5. Signature: _____   6. Date:_____

7. Phone Number: (____)_____   8. SIC Code:_____

9. Fax Number: (_____)_____

LISTING INFORMATION:

(1) CLASSIFICATION:
(Review all first, then select the **one** that best describes your firm's service)

- ❏ Recycling
- ❏ Equipment and Supplies
- ❏ Environmental Consulting
- ❏ Legal and Health - Related Services
- ❏ Tank Management
- ❏ Collection and Transportation
- ❏ Storage, Treatment, Disposal
- ❏ Emergcy Response/Clean-Up
- ❏ Laboratory Analysis
- ❏ Waste Minimization
- ❏ Miscellaneous

(2) DESCRIPTION OF SERVICE:
(In 25 words or less, please describe your firm's service or product, keeping in mind what the reader of your listing may want to know. You may use the space provided below or type the listing on a separate piece of paper and attach it to this form.)

(3) LOCATION SERVED:
(Give general area where service is available, e.g., Central Georgia, Southeastern U.S., etc.)

Send completed form to: SWIX Clearinghouse
Post Office Box 960
Tallahassee, FL 32302

Figure 4.1. SWIX Listing Form

These waste exchanges provide the connection between the waste generator and waste user. You may notice that only the physical separation of the generator from the user differs from the solution proposed in Chapter 5. Waste exchanges can be of either the informational or the material type.

Material Available/Wanted
Listing Form

A separate form is required for each item listed.
Limit of two listings.

1. Company Name:_____ SIC Code #:_____

2. Mailing Address:_____

   _____

3. Company Contact:_____ 4. Title:_____

5. Signature:_____ 6. Date:_____ 7. Phone: (___)_____

8. Fax Number: (___)_____

9. Check One Only: ❏ MATERIAL AVAILABLE   ❏ MATERIAL WANTED

   Classifications (review all first; then select one that best describes your material):

   ❏ ACIDS   ❏ OTHER ORGANIC CHEMICALS   ❏ WOOD AND PAPER
   ❏ ALKALIS   ❏ OILS AND WAXES   ❏ METALS AND
   ❏ OTHER INORGANIC CHEMICALS   ❏ PLASTICS AND RUBBER   METAL SLUDGES
   ❏ SOLVENTS   ❏ TEXTILES AND LEATHER   ❏ MISCELLANEOUS

10. Material to be listed (Main usable constituent, generic name):_____

11. The industrial process that generates this waste:_____

12. Main constituent and percentage:_____

13. Other constituents (including contaminants):_____

14. Percent by (check one):   ❏ Volume   ❏ Wet Weight   ❏ Dry Weight

15. Physical State: ❏ Solution   ❏ Slurry   ❏ Sludge   ❏ Cake
                      ❏ Aggregate ❏ Solid   ❏ Dust   ❏ Gas

16. Miscellaneous information (e.g. pH, toxicity, reactivity, color, particle size, flash point, total solids, purchase date, manufacturer):

    _____

    _____

17. Potential or intended use:_____

18. Packaging: ❏ Bulk   ❏ Drums   ❏ Pallets   ❏ Bales   ❏ Other:_____

19. Present Amount:_____   20. Frequency: ❏ Continuous   ❏ Variable   ❏ One Time

21. Quantity thereafter:_____   ❏ Pounds   ❏ Tons   ❏ Cubic Yards   ❏ Gallons
                         ❏ Kilograms ❏ Cubic Meters ❏ Liters   ❏ Other_____

    per: ❏ Day   ❏ Week   ❏ Month   ❏ Quarter   ❏ Year

22. Restrictions on amounts: ❏ None   ❏ Minimum   ❏ Maximum_____

23. Available to Interested Parties: ❏ Sample   ❏ Lab Analysis   ❏ Independent Analysis

24. For material wanted, acceptable geographic area (i.e. States, regions, countries):

    _____

25. If necessary to speed communications, please check if your company's name, address and telephone number may be released.   ❏ YES   ❏ NO

Send completed form to: **SWIX Clearinghouse**
                                 Post Office Box 960, Tallahassee, FL   32302

Figure 4.2. Participation in SWIX

In a 1985 symposium on management of toxic and hazardous wastes, Simpson described the 1981 law creating the New York State Industrial Materials Recycling Act.[10] This Act mandated the creation of a public benefit entity known as the New York State Environmental Facilities Corporation (EFC). One of the objectives of the EFC is to encourage the exchange of materials. Simpson differentiates the "active" from the "passive" exchange. In the latter type, generators list wastes they wish to transfer and potential users list wastes they desire. The information in catalog or brochure form includes the quantity, description, availability, and general location. Those interested in particular wastes may contact a waste exchange, which will forward their inquiries to the lister, who responds to the inquiries. Exchanges of this type, is similar to the Florida SWIX exchange, do not become involved in any subsequent negotiations.

On the other hand, an "active" type of exchange acts as an intermediary between the generator and the user. In the New York State active system, the EFC staff has concentrated on making direct contact with industry via surveys of small businesses. Simpson reports several successful exchanges in the first years of exchange operation such as ammoniated copper etching, NH_4OH, $Cu(OH)_2$, $FeCl_3$, NH_4Cl, TCE, alcohol, benzene, oil, acetylene, calcium hydroxide, coal, paint, perchloroethylene, glycerine, and $Ca(OH)_2$. He also gives broader powers of the EFC including financing pollution control facilities for private industries with attendant tax exempt securities. The State's Industrial Materials Recycling Act includes provisions for the establishment of procedures to protect trade secrets and other proprietary information pertinent to the

---

[10]Simpson, P.T. "Management of Toxic and Hazardous Wastes," *New York State Industrial Materials Recycling Program*, ed. H.G. Bhatt, R. Sykes, and T. Sweeney (Chelsea, Mich.: Lewis Publishing, 1985), chap. 16, p. 195.

industry's operation. Failure to keep such information confidential subjects an employee to dismissal or a fine up to $5,000.

In the same symposium Banning revealed the presence of three major multistate exchanges; The Northeast Industrial Waste Exchange, The Industrial Material Exchange Service, The Midwest Industrial Waste Exchange, in addition to the aforementioned Southern Waste Information Exchange.[11] He also points out specific state exchanges that exist in California, Georgia, and New Jersey, as well as local city exchanges such those in Louisville and Houston.

Banning acknowledges that there are many impediments to successful use of the industrial waste exchange system. It is important, however, to be aware of the problems that exist. One difficulty is that many small industrial plants have only a small amount of waste available on an unpredictable, infrequent basis. A reuser or purchaser must collect these small quantities of waste over a considerable geographical area.

In addition, similar wastes may not be compatable because they contain different minor contaminants. Another problem for industrial waste generators is caused by the provision of RCRA (Resource Conservation and Recovery Act), which requires them to obtain a permit if they store the small quantity of waste for more than 90 days. Both obtaining the permit and storing enough material for a buyer are difficult and cumbersome. Still another impediment to waste exchanges are the "restrictive, uncertain, misunderstood, or feared hazardous waste management regulations existing at the state and federal

---

[11]Banning, W. "The Role of a Waste Exchange in Industrial Waste Management—Development of the Northeast Industrial Waste Exchange," New York State Industrial Materials Recycling Program, ed. H.G. Bhatt, R. Sykes, and T. Sweeney (Chelsea, Mich.: Lewis Publishing, 1985).

levels." It is hoped that during the last 10 years some of these issues have been clarified. The inconsistency among state regulations has hindered interstate transfers. One problem, the identification of markets of recycling opportunities, has been alleviated somewhat by more widely disseminated information. However, Banning feels that we should focus our attention on the problems faced by small, widely dispersed generators of industrial waste. Once again, this could be accomplished by using the EBIC (Environmental Balanced Industrial Concept) concept that I propose in Chapter 5.

Crepeau and Beltz, in the same 1985 symposium, present the European Network of Waste Exchanges, which has existed in Europe since 1969.[12] They describe a unique arrangement in which technological or research institutes maintain waste exchanges in Denmark, Norway, Sweden, Finland, and Iceland. This exchange, known as the Nordic Waste Exchange, is operated by the Swedish Water and Air Pollution Research Laboratory and is funded jointly by the Swedish government and industry.

The EPA, in its Bibliographic Series[13] describes the art of waste exchange as a means of recycling hazardous wastes. This publication includes an annotated section on waste exchange publications during the 1980s, as well as a list of specific exchanges giving names of contact people, telephone numbers, and addresses. The reader is encouraged to review these sections for more details on this subject.

In a 1987 *Technical Resource Document* on treatment technologies for halogenated organic wastes, the EPA con-

---

[12]T. E. Crepeaw and P. E. Beltz. "European Network of Wast Exchanges," *New York State Industrial Materials Recycling Program*, ed. H.G. Bhatt, R. Sykes, and T. Sweeney (Chelsea, Mich.: Lewis Publishing, 1985).

[13]*Waste Minimization: Hazardous and Non-Hazardous Solid Waste (1980 to Present)*, U.S.E.P.A., EPA/IMSD - 87 - 007, September 1987. Washington D.C.: U.S. Government Printing Office.

siders the "exchangeability" of a waste to be significant.[14] Such material is enhanced by higher concentration and purity, quantity, availability, and higher offsetting disposal costs. Some of the limitations to waste exchangeability are the high costs and other difficulties associated with transportation and handling, costs of required purification or pretreatment, and, in certain cases, the effect on process or product confidentiality.

[14]Environmental Protection Agency, *Technical Resource Document Treatment Technologies for Halogenated Organic Containing Waste* (Cincinnati, Ohio: U.S. Environmental Protection Agency, Section 4, p. 9.

CHAPTER 5

ENVIRONMENTALLY BALANCED INDUSTRIAL COMPLEXES: AN INNOVATIVE SOLUTION TO ATTAIN ZERO POLLUTION

A—Introduction
B—Pulp and Papermill Complexes
C—Tannery Complexes
D—Sugarcane Complexes
E—Textile Complexes
F—Fertilizer–Cement Complexes
G—Fossil Fuel Power Plant Complexes
H—Steel Mill–Fertilizer–Cement Complexes
I—Plastic Manufacturing Industrial Complexes
J—Cement–Lime–Power Plant Complexes
K—Wood (Lumber) Mill Complexes
L—Power Plant–Agriculture Complex
M—Power Plant–Cement–Concrete Block Complexes
N—Cannery–Agriculture Complex
O—Nuclear Power–Glass Block Complex
P—Animal Feedlot–Plant Food Complex

INTRODUCTION

In their 1993 approach, Schiller and Hackman approach the subject of and solution to "zero discharge" by describing it not as pollution elimination, but as the "lowering of levels of discharge by products of human activities to values comparable to existing backround levels."[1] They do not believe that currently such an approach is economically and technically unfeasible. However, from a practical point of view they suggest that every manufacturing process, from agricultural activities to nuclear power production, should be reexamined to identify ways and means by which current technology can alleviate generation and discharge of pollutants. They point out the very obvious, but often overlooked, result of waste treatment: that contaminants removed by treatment must be handled. And the solution, naturally, is to reuse the removed contami-

---

[1] M. Schiller and M. Hackman, "Water Re-use Systems for Zero Discharge" *Environmental Protection* (September 1993): 72–73.

the solution, naturally, is to reuse the removed contaminants as raw materials in some manufacturing process. In addition, reuse of a waste "is not only process specific, but also area specific and scale dependent. "They qualify this by stating that "what may be feasible in a highly industrial area might not be feasible in a rural area." Likewise, a method applicable to a 10-ton-per-day system might not be feasible for a 100-pounds-per-day system, and vice versa. They caution that while it may be possible to attain "zero discharge" with the water discharge alone if the removed solids are not reused, true total zero discharge has not been achieved. Spills, process upsets, and periodic "blowdowns" of solids also prevent the attainment of true zero discharge. However, the benefit of eliminating permit violation fines is significant. They conclude that "since there is no treatment without residual generation, residual management, including disposal, constitutes a substantial part of all wastewater treatment costs." This is the very reason that this chapter concentrates on reuse directly, without treatment, to avoid these costs.

The absolute optimum in waste minimization involves developing a system whereby no wastes from industry are discharged into the air or water, or onto the land. This is rather difficult to accomplish for an individual industrial plant at economically feasible levels. It would require either (1) complete treatment of all wastes, with subsequent reuse of all treated effluents, or (2) exportation of all wastes in treated or untreated states to some distant user or disposer. Both alternatives are expensive and/or difficult to achieve. Deductive reasoning leads to proposing an ideal system for solving industry's dilemma.

The system we propose is that of combining, in one complex, two or more complementary industrial plants, to manufacture products in tandem so that the wastes of one serve as the raw materials of the other. I have referred to such complexes since 1977 as Environmentally Balanced Industrial Complexes (EBICs).[2]

Environmentally Balanced Industrial Complexes are simply a selective collection of compatible industrial plants located together in one area (complex) to minimize both environmental impact and industrial production costs. These goals are accomplished by utilizing the waste materials of one plant as the raw materials for another with a minimum of transportation, storage, and raw material preparation. When a manufacturing plant neither treats its wastes, nor stores or pretreats certain of its raw materials, its overall production costs must be reduced significantly.

Large water-consuming and waste-producing industrial plants are ideally suited for location in such industrial complexes. Their wastes, which may cause pollution if released to the environment, may be amenable to reuse by close association with satellite industrial plants using the wastes and producing raw materials for others within the complex.

Elimination of waste treatment costs alone may be sufficient to influence industrial managements to continue to produce their products in the highly competitive world market. Although environmental engineers may not be involved in this decision, it should be their goal to minimize waste treatment costs and maximize protection of the environment.

It may often be difficult to identify any reuse costs within such complexes, but they should be more easily absorbed into production costs than end-of-the line waste treatment costs. Although the advantages of these complexes are obvious, certain questions must be answered

---

[2]N. L. Nemerow, S. Farroq, and S. Sengupta, "Industrial Complexes and Their Relevance for Pulp and Paper Mills," *Proceedings of the Seminar on Industrial Wastes*, Calcutta, India, December 8, 1977.

and provisions made to ensure smooth and effective multioperational production. For example, waste quantities and raw material needs must be ascertained and matched. Questions must be answered as to the type of labor and number of workers available in an area, the feasibility of marketing of several products from one location, and the local tax levels for varying types of plants. Optimal mass balances require identification of amounts and types of each product and the economics involved with manufacturing each of them.

In the last 13 years, I have either worked on, proposed, or been familiar with many potential compatible industrial complexes, which include the following industries:

1. Pulp and Paper Mill
2. Tannery
3. Sugarcane
4. Textile
5. Fertilizer–Cement
6. Fossil Fuel Power Plant
7. Steel Mill–Fertilizer–Cement
8. Plastic
9. Cement–Lime and Power Plant
10. Lumber Mill

There are many other potential complexes, as well as variations of these combinations. The key to feasibility for any complex lies finally in production economics and environmental protection. The following sections of this chapter discuss in some detail the 10 complexes listed, as working illustrations of the system. These should serve as examples for others to use or modify as desired in real situations.

Solid waste is only a misplaced raw material. Someone or some industry somewhere can use the misplaced raw material to make a valuable product. The problem arises in the matching of the waste producer with the raw material user. Once we locate the producer and user, we must resolve other questions before the "marriage" can materialize. The questions involved are (1) *distance* (physical) between the two, (2) *economics* of joining the two plants, and (3) *compatibility* of the waste material from one plant as a raw material for the other.

Distance

Industrial plants, up to now, have considered locating where factors such as raw materials, labor, market, taxes, utilities, community cooperation, transportation access, and so forth, are favorable. This practice may place our suitors many miles apart, since responses to these factors often differ from industry to industry. How can we get them to locate adjacent to each other? Obviously, we must prove that it is in the best interest of two compatible plants to locate near each other in the same complex. An industry that does not have to treat its wastes nor transport or store its raw materials will experience lowered production costs. At the same time, no adverse environmental impact will result, since no waste materials will leave the complex.

Economics

Industry wants and deserves facts that prove it is less expensive to manufacture both plants' products in the same complex than to do this at locations distant from each other. The only way to provide this verification is to do prototype analysis, generate real data, and feed the data into an analytical model to yield final production costs. The ultimate solution of the optimum model is one which gives the lowest production costs. A key element in determining the optimum solution is that of including the real environmental costs of avoiding pollution of air, water, and land as part of production costs. These costs must be included in the analytical model.

Compatibility

The two (or more) plants included in an EBIC must be compatible. Mainly, this refers to the acceptability of one plant's waste (or wastes) as raw material(s) for the other's production. Such waste will usually substitute for some or all of one raw material. In that case, the waste may differ slightly from the usual raw material. This may mean that the receiving plant may have to accept somewhat altered manufacturing techniques or slightly altered product specifications. Sometimes wastes may require some modification prior to use as raw material. This possibility should be considered in figuring production cost inputs for the analytical model.

Compatibility also refers to the cooperative abilities of the plants within the complex. Each plant becomes dependent upon the other(s). Factors that affect the production of one plant will affect its wastes, either by volume or by character, and these in turn will affect the production of the other plant(s). Machine or utility breakdowns, fires and other catastrophes, changed market conditions, temporary labor strife, and so forth, will all affect production schedules and the availability of wastes, which must be considered by the ancillary plant(s) in the complex. These situations arise when cooperating plants are at some distance from each other, such as when the delivery of raw materials is delayed. In fact, it may be easier to ameliorate these factors within a complex than at plants distant from each other. In any event, all plants must be cooperative, understanding, and flexible enough to operate when unusual situations develop.

The ideal solution to waste problems of all kinds, gaseous, water, and solid, is the ultimate recycling or complete reuse within an industrial complex. This produces zero pollution, affects no environment external to the complex, and results in lowered production costs for the industrial plant's and possibly greater benefits. Chapter 15, *Industrial and Hazardous Wastes*, describes briefly some groupings of compatible industrial plants and show how the wastes produced are used by the receiving plants.[3] However, most of these complexes are only theoretical at this time and are not engaged in actual production, but indications are that some are getting closer to implementation.

An argument set forth by those supporting the status quo or opposing recovery and reuse as a solution to environmental pollution is that markets for recovered wastes are unstable. They cite examples of markets for paper, glass, and aluminum that often disappear or decrease radically when economic conditions deteriorate. The same may be true for wastes recovered within industrial complexes.

I believe that, while these facts may be partially valid, there are also compelling counter arguments. If the price of potential environmental damage is subtracted from the cost of the recovered waste product, the price of the resulting raw material would be competitive under any market conditions. In fact, a declining economy should enhance, rather than decrease, the use of recovery and reuse as an optimum pollution abatement technique.

There are a great variety of plants that can be combined in an industrial complex. The following sections of this chapter discuss in detail ten types of EBICs.

---

[3] N. L. Nemerow and A. Dasgupta *Industrial and Hazardous Wastes* (New York: Van Nostrand Reinhold, 1991).

PULP AND PAPER MILL COMPLEXES

The products of pulp and paper mills, the fifth largest industry in the United States economy, are consumed at the annual rate of 400 pounds per person.[4] As early as 1976, the U.S. production of paper and paperboard was about 64 million tons. At that time, more than half the world's production was consumed in the United States.[5]

Wood and other raw materials are pulped mostly by chemical means, primarily by alkaline sulphate or the kraft process. Then the various pulps are converted into paper and paperboard products. The pulping of wood and the formation of paper products evolves wastes that contain sulfates, fine pulp solids (fillers), bleaching chemicals, mercaptans, sodium sulfides, carbonates and hydroxides, sizing casein and resins, clay, inks, dyes, waxes, grease, oils, and other small fibers.

The overall (mixed) waste varies from a low to a high pH level, depending on the effluent being discharged at the moment, and contains high-color, suspended, colloidal, and dissolved solids and inorganic fibers. Because of its high water consumption and subsequent wastewater discharge of 20,000 to 60,000 gallons per ton of product, the waste contains large total quantities of organic, oxygen-demanding matter. For a complete analysis of these wastes, see Ref.4, page 113.

The high water use and wastewater production usually preclude the possibility of joint treatment with municipal sewage. These wastes also create considerable environmental impacts because of their concentrated loads of air, water, and land pollutants. The siting of new pulp and paper mills today has become a major endeavor. Proximity to cutting forests and market outlets are not the only con-

---

[4] Ibid., pp. 461–476.
[5] R. N. Steve and J.A. Brink, Jr., *Chemical Process Industries*, 4th ed. (New York: McGraw-Hill, 1977), 556.

siderations. These mills must be located near large quantities of relatively clean water, as well as receiving water resources, downwind and at a considerable distance from residences (because of common air pollutants such as SO_2 and mercaptans), usually on a rail line and near major highways for shipping, and close to adequate land area for conventional waste treatment (if used) and sewage disposal. Sites with these many features are difficult to find. For this and other reasons previously mentioned, I recommend using an Environmentally Balanced Industrial Complex.

Pulp and paper mills are also among the five major industrial water users in the United States, consuming about 2×10^{12} gallons per day (7.5×10^6 m³/day) according to M. Gould.[6] Most of this is discharged into the environment as wastewater from the washing process or as steam from the drying plant. Nearly 50% of the wood entering the pulp mill leaves the mill as product paper. Good and Trocino[7] report that 14 million tons of bark removed each year have become a disposal problem since uses of fibers as fuel have been outstripped by production.

The greatest percentage of loss in weight from trees to paper is represented by solids, which must be disposed of into the environment. Of these, bark, waste pulp, and paper mill fines constitute the majority and end up in the land or the air.

Because of the significance of this industry's wastes to the environment and because of the added cost of waste treatment required for conventional solutions, we have derived a pulp and paper mill complex for further investi-

---

[6]M. Gould, "Water Pollution Control in the Paper and Allied Products Industry", *Industrial Wastewater Management Handbook*, (New York: McGraw-Hill, 1976)

[7]R. D. Good and F. S. Trocino, "Fir Bark Conversion Route," *Process Technology and Flow Sheets* (New York: McGraw Hill, 1979), 310.

gation. The 1,000 tons per day of fine paper product is the center of the complex proposed in Figure 5.1.

We presented a balanced industrial complex centered about a pulp and paper mill as early as 1977.[8] Eight separate industrial plants were included, five of which would produce products to be used within the complex.

Timber is brought into the complex to the pulp mill (1) shown in Figure 5.1. Major wastes from (1) are bark, which is burned subsequently in the steam plant, and sulfate waste liquor, which is used in three internal plants: road binder (3), vanillin (4), and sulfate concentrating (8). Products from (3) and (4) can be sold locally or internationally, and those from (8) are used within the complex by (1) or by the hardboard manufacturing plant (7). Fine paper product from (2) can be sold in the world market. Wastes from (2) include heat, fillers, and fines that can be used internally in the groundwood pulp mill (5), which also uses a percentage of used and recycled newspaper stock. The pulp product from (5) will be used partially within the complex by (1) and also sold as paperboard externally. The plant (5) produces waste of suspended solids, which are used externally by the wrapping paper plant (6) and the pressed hardboard plant (7). The products of (6) and (7) can be sold regionally. In all, this complex manufactures six products for external sale: fine paper, wrapping paper, hardboard, vanillin, paperboard, and road binder. It also produces four products for internal use: concentrated sulfate, wood pulp, wrapping paper, and groundwood pulp. In addition, all the major wastes of suspended solids, cooking liquor, fillers, heat, and bark are reused within the complex in the manufacturing of these products.

[8]Nemerow, Farooq, and Sengupta, "Industrial Complexes and Their Relevances for Pulp and Papermills," Vol. 3, No. 1, page 133 Dec. 8, 1977 Calcutta, India. Published in 1980 by Pergamon Press, Oxford, England.

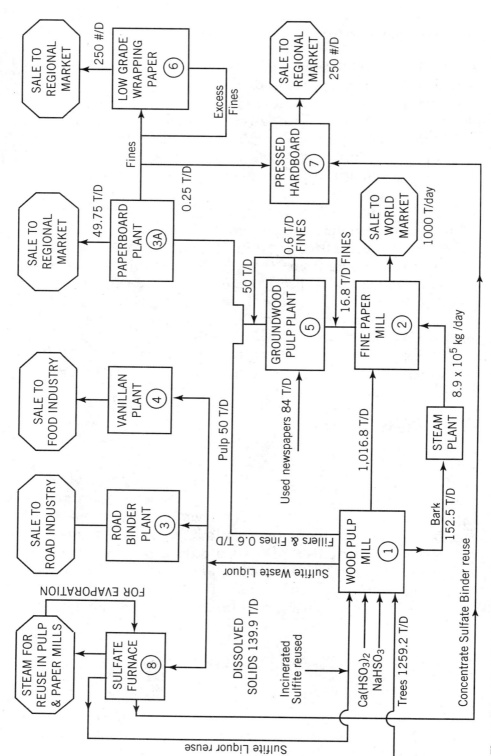

Figure 5.1. Pulp and Paper Mill Complex

Mass Balance of Products

A literature review revealed typical concentrations of recoverable suspended solids in various process effluents.[9] A mass balance was prepared assuming that the daily production of fine paper is 1,000 tons. The remaining quantities are calculated based on this production.

Computation of daily weight of trees required at the complex: Production of fine paper = 1,000 tons per day (907.2 kg/day $\times 10^3$).

Fiber losses from paper mill = 1.68% of production. Therefore, suspended solids going into waste streams from the paper mill = 1.68/100 $\times 10^3$ = 16.8 tons/day (15.24 kg/day $\times 10^3$).

Total groundwood pulp produced and lost per day = 100 + 0.6 = 100.6 tons/day (912.6 $\times 10^2$ kg/day). Therefore, raw material (used newspaper) required = 100.6 − 16.8 = 83.6 tons/day (758.4 $\times 10^2$ kg/day).

Sulphate Recovery

The solids concentration of spent sulfate liquor waste from the pulp mill (1) digester will vary from 6% to 16%, with an average value of 11%. These solids may contain as much as 68% liquosulfonic acid, 20% reducing sugars, and 6.8% calcium. (6,'77) Complete evaporation of sulfate waste liquor produces a fuel that can either be burned without an additional outside fuel supply or yield a salable by-product such as synthetic vanillin (4) or road binder (3). The overall potential production of kraft papers of different quality is presented in Figure 5.2.

---

[9] R. N. Tewari, and N. L. Nemerow, "Environmentally Balanced and Resource Optimized Kraft Pulp and Papermill Complex," Proceedings of the 37th Purdue Industrial Waste Conference, Purdue University, Lafayette, Ind., May 12, 1982, p. 353.

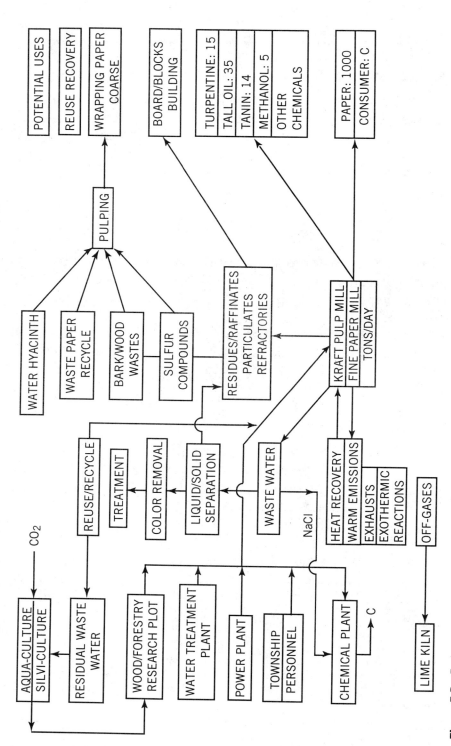

Figure 5.2. Environmentally Balanced Pulp and Paper Complex

In recovering all the suspended matter and most of the wastewater, it must be presumed that a considerable portion of dissolved and colloidal organic matter is being incorporated into the various products. This is especially true in the case of sulfite waste liquor, which is completely reused or recovered and contains most of the BOD in the complex. If we presume that 50 % of the groundwood pulp is recycled (as shown in Figure 5.1), the rest is used to produce paperboard (3A).

Paperboard Production

Since the loss as fines from groundwood pulp production (5) is about 0.5% of production,[10] and if x is equal to the paperboard (Figure 5.1, 3A) production per day, then

$$x + 0.15x = 50 \text{ tons of pulp/day } (543.75 \times 10^2 \text{ kg/day}),$$

$$x = 49.75 \text{ tons paperboard/day } (451.32 \times 10^2 \text{ kg/day}), \text{ and}$$

$$\text{Recovered fines from paperboard waste} = 50 - 49.75 = \frac{0.25 \text{ tons/day}}{(226.79 \text{ kg/day})}.$$

Now, 0.25 tons of fines per day can be used to produce low-grade wrapping paper and pressed hardboard (6 and 7). With no loss of fines and with a 50 - 50 production split, 250 pounds (113.64 kg) of each product can be manufactured.

Energy Management

Integrated production complexes such as shown in Figure 5.1 have, from the standpoint of energy management, a significant advantage over conventional plants located separately. Waste heat from one plant can be used by

---

[10]N. L. Nemerow, *Industrial Water Pollution: Theories, Characteristics and Treatment* (Reading, Mass.: Addison Wesley, 1978), 443.

another within the complex. Environmental problems from waste-heat discharges are well known, such as thermal discharge waters causing stratification in receiving waters, lowering oxygen content, and increasing biological metabolic rates.

This pulp and paper mill complex will reduce waste heat discharged to the hydrosphere and atmosphere by:

1. Utilization of solid waste
2. Utilization of low grade heat within the complex

Solid Wastes

Bark from the tree-shredding plant is used as fuel the steam plant to provide heat in the papermill (2). Bark, in the amount of 152.5 tons per day, with an average heating value of 4,000 calories per gram[11] yields:

$$(\frac{cal}{gm}) \, 152.5 \times 2000 \frac{lb}{ton} \times 4kcal \times 454 \frac{gm}{lb} =$$

$$5.53 \times 10^8 \frac{kcal}{day} (2.3 \times 10^{12} joules) \times$$

$$454 \frac{gm}{lb} (2.3 \times 10^{12} joules).$$

At an incoming water temperature of 25°C, the steam produced is

$$(5.53 \times 10^8) \, / \, (75 + 540) = 8.9 \times 10^5 \text{ kg/day.}$$

Evaporation and then burning of sulfite waste liquor can yield an energy source within the complex (Figure 5.1, plant (8). In this proposed complex the manager has choices (as shown in Figure 5.1) to reuse concentrated

[11]Tewari and Nemerow, p. 353.

liquor in the pulp mill (1) or in making hardboard (7), or to burn it completely to make steam for reuse mainly in the evaporator room (8) or in the papermill (2). Operating difficulties often arise from the scaling, corrosion, and fly ash in the evaporators and boilers. However, burning may also be justified by eliminating need to discharge highly polluted sulfite waste liquor to the exterior environment.

Utilization of low-grade waste heat from the proposed complex is somewhat difficult to quantify without further process thermodynamic analysis. However, some low-grade waste heat (from the paper roll driers) may be used in the groundwood pulp mill (to operate and drive grinders, etc.) In colder climates, waste heat from many of the effluents can be used for space heating and to provide hot water for plant personnel.

Economical Benefits Associated with This Potential Complex

In addition to elimination of any costly wastewater treatment plants, little or no air, water, or land pollution results from operations of the complex. By eliminating waste treatment costs alone, a savings of 1% to 5% of production costs can be realized.[12] Further savings will be realized in the following areas:

1. Reusing burned sulfite waste liquor to replace a portion of calcium or sodium bisulfite cooling liquor
2. Reusing 50 tons per day of groundwood pulp (5) (43.359×10^3 kg/day) to replace a similar weight (or more) of trees in the wood pulp mill (1)
3. Burning 152.5 tons/day of bark from (1) to use the steam in the fine paper mill (2)
4. Reusing concentrated sulfite waste liquor from the evaporators in (8) to make pressed hardboard (7)

[12]Nemerow and Dasgupta, *Industrial and Hazardous Wastes*, pp. 81–98.

5. Reusing 16.8 tons per day of fillers and fines (along with the heat) from the paper mill (2) to make groundwood pulp (5)
6. Reusing 0.6 tons per day of groundwood pulp mill (5) fines (544.31 kg/day) to make additional pulp in (1) or paperboard in (3A)
7. Reusing 0.25 tons per day of paperboard mill fines (226.80 kg/day) from (3A) to make both low-grade wrapping paper in (6) and pressed hardboard in (7)
8. The sale of additional products outside the complex:
   250 pounds per day (113.64 kg/day) of low-grade wrapping paper from (6)
   250 pounds of pressed hardboard (113.64 kg/day) from (7)
   49.75 tons per day (45.132 x 103 kg/day) of paperboard from (3A)
   Undetermined amount of vanillin per day from (4)
   Undetermined amount of road binder per day from (3)
9. Complete burning of a concentrated (by evaporation) undetermined amount of sulfite waste liquor to recover heat and abate environmental pollution from (8)
10. Use of low-grade waste heat from various plants to be used for space heating and water heating

Total wood pulp produced per day is

$$1000 + 16.8 = 1{,}016.8 \text{ tons } (922.44 \text{ kg} \times 10^3)$$

Quantity of sulfite liquor generated
in the wood pulp mill =
300 gallons/ton ($1.24 \text{ kg} \times 10^{-3}$) of pulp produced
at a concentration of 11% in sulfite liquor.

Thus, the dissolved solids going into sulfite liquor =
275.22 lb/ton =
110,000 lb/m $\times$ 8.34 $\times$ lb/gal $\times$ 300 gal $\times$ 10^{-6}m lb.

Total sulfite waste water dissolved solids
produced per day =
275.22 lb/ton $\times$ 1016.8 tons/day $\times$ 1 ton/2000lb =
139.9 tons/day (1.269 $\times$ 10^{-5} kg/day).

On the assumption that the amount of *bark produced* is
generally 15% (by weight) of the pulp production,

The bark produced = .15 $\times$ 1,016.8 tons/day =
152.5 tons/day (or 1.38 $\times$ 10^5 kg/day).

Total tonnage of trees used in the complex [Plant (1)] =
1,016.8 + 139.9 + 152.5(bark) =
1,309 tons/day (1.187 kg $\times$ 10^6/d) paper sulfite +
fines waste solids

As to groundwood pulp production [Plant (5)], 16.8
tons/day (1.52 $\times$ 10^4 kg/day) of fines are recovered from the
papermill. If 100 tons (907.2 kg $\times$ 10^2) of ground pulp is
required for production each day, then 0.6 tons will be lost
as fiber each day (544.3 kg/day).

TANNERY COMPLEXES

Tanneries convert animal skin into leather for many
apparel uses. The conversion takes place in two major
parts of the plant: the beamhouse-and the tanhouse. In
the *beamhouse*, the processes consists of curing, fleshing,
washing, soaking, dehairing, lime splitting, bating, pickling, and degreasing. In the *tanhouse*, the final leather is

prepared by vegetable or chrome tanning, shaving, and finishing. The latter process includes bleaching, stuffing and fat-liquoring, and coloring. A detailed schematic presentation of both major operations is shown in Figure 5.3.

The predominant and most significant pollutional waste from these plants originates in the upper-sole chrome-tanning mills. This waste is hot, highly alkaline, odorous, highly colored, and contains large quantities of dissolved organic matter, BOD, suspended solids, lime, sulfides, and chromium. An overall total tannery waste averages about 10,000 gallons of wastewater per 1,000 pounds of wet, salted hide processed. The waste contains an average of 8,000 ppm total solids, 1,500 ppm volatile (organic) solids, 1,000 ppm protein, 300 ppm NaCl, 1,600 ppm total hardness, 1,000 ppm sulfide, 40 ppm chromium, 60 ppm ammonium nitrogen, and 1,000 ppm BOD.[13] It has a pH between 11 and 12 and normally produces a 5% to 10% sludge concentration (when settled) because of the lime and sodium sulfide contents.

The treatment of such wastes had been difficult because of the conflicting pollutional parameters of high pH, high organic matter content, and potential toxic compounds. Most successful treatment plants today use some form of biological treatment (as contrasted with chemical treatment in the 1950–1970 era) to reduce the oxygen demand on receiving waters. This necessitates the use of well designed and operated preliminary treatments to ensure safe and efficient biodegradation. High sludge quantities result from these treatments. Therefore, properly designed and operated tannery waste treatment systems may be costly to build and operate, yet the absence of these facilities will result in excessive stream pollution. Locating tanneries in environmentally balanced industrial complexes eliminates both of these negatives.

---

[13]N. L. Nemerow, *Industrial Water Pollution*, 335.

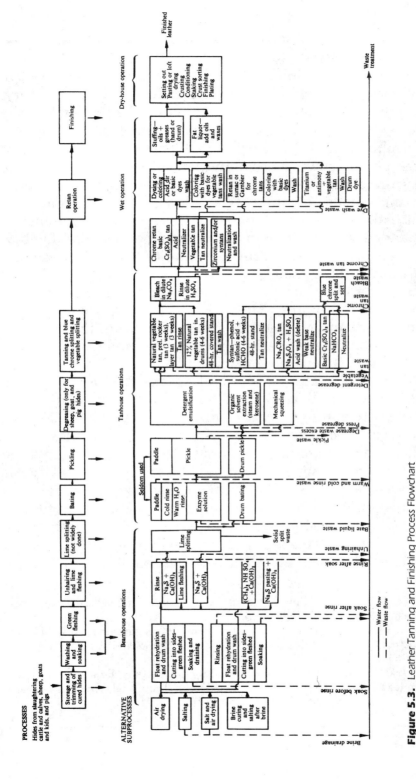

Figure 5.3. Leather Tanning and Finishing Process Flowchart

Source: N. L. Nemerow, *Industrial Water Pollution: Origin, Characteristics, and Treatment* (Melbourne, Fla.: Krieger Publishing, 1987).

To derive a feasible complex that includes a tannery, we must select industrial plants compatible with the tannery. It stands to reason that the raw material supplier and the market outlet user of the tannery represent such industrial plants. In 1977, 1980, and 1981 I reported on a first attempt at such a combination of plants.[14,15,16] The fulcrum plant of this complex is the tannery; they supporting compatible industrial plants are the slaughterhouse and rendering plants. The three-industry complex is expanded to include an animal grazing and feedlot facility, as well as a residential community for all personnel working in the complex.

Biogas and power plant services make the complex even more self-sustaining. Outside service requirements are minimized, and all power is generated within the complex. Excess products of leather, meat, soap, and even electricity, are sold to outside consumers.

Chemicals, water, cattle, and animal feed are imported to the complex. Wastewater, blood and bonemeal, hide and leather trimmings, cattle dung and residential solid wastes are recovered and reused within the complex.

The mass balance centered about a medium-sized tannery producing 36,000 square feet of finished leather is shown in Figure 5.4. Residential as well as industrial solid wastes are fermented to methane gas, which is used subsequently to produce electrical energy for use in the complex. Waste sludge from the fermenter is incinerated to produce additional electrical energy. External raw materi-

---

[14]N. L. Nemerow, "Environmentally Optimized Industrial Complexes," *Proceedings of NEERI*, Nagpur, India, 1980.

[15]N. L. Nemerow, "Preliminary Assessment of Environmentally Balanced Industrial Complex: Three Stage Evolution," *Report to U.S.E.P.A.* Contact No. 68-02-3170, RTP, North Carolina June 1980.

[16]N. L. Nemerow, and A. Dasgupta, "Environmentally Balanced Industrial Complexes" 36th Annual Purdue University, Industrial Waste Conference Lafayette, Indiana, 1981 Proceedings, p. 416.

Table 5.1 External Raw Materials and Manufactured Products in Three Industry Complex (STAGE 3)

	Raw Material Required from Outside the Complex		Manufactured Products for Outside Sale	
	Material	Amount	Material	Amount
1.	Fresh Makeup water	2,927,599 gal/d	1. Meat products	513,341 #/d
1A.	Well water (one time only)	12 MGD	2. Tanned leather	36,000 sq.ft./d
			3. Tallow	79,740 #/d
2.	Calves	900/d (150 days) 540,00 #/d	4. Energy	694,710 KWH/d
3.	Chemicals	495 #/d Na_2S 3960 #/d $Ca(OH)_2$ 1500 #/d H_2SO_4 2475 gal/d kerosene 1980 #/d oil or wax 2475 #/d $Cr_2(SO_4)_3$ 4208 #/d NaCl 7 #/d Cl_2		
4.	Cattlefeed	2,625,000 #/d		

Source: Nemerow and Dasgupta, *Industrial and Hazardous Wastes*, p. 209.

als required and manufactured products for external sale are listed in Table 5.1.

The killing, dressing, and some by-product processing of animals are carried out in the slaughterhouse. This plant is the supplier of raw materials, hides for the tannery, as shown in Figure 5.4, while producing its main product for sale externally, fresh meat carcasses. The animals are stuck and bled on the killing floor. Carcasses are trimmed, washed, and hung in cooling rooms.

Livers, hearts, kidneys, tongues, brains, and so forth are, sent to cooling rooms to be diluted before being marketed. Hides, skins, and pelts are removed from the animals, sorted, and placed in piles until they are shipped to tanneries or wool-processing plants. In our proposed

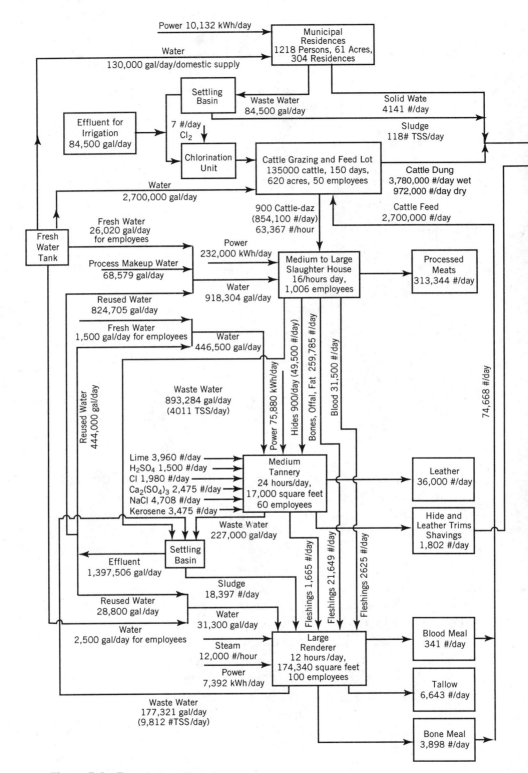

Figure 5.4. Three Industry Complex

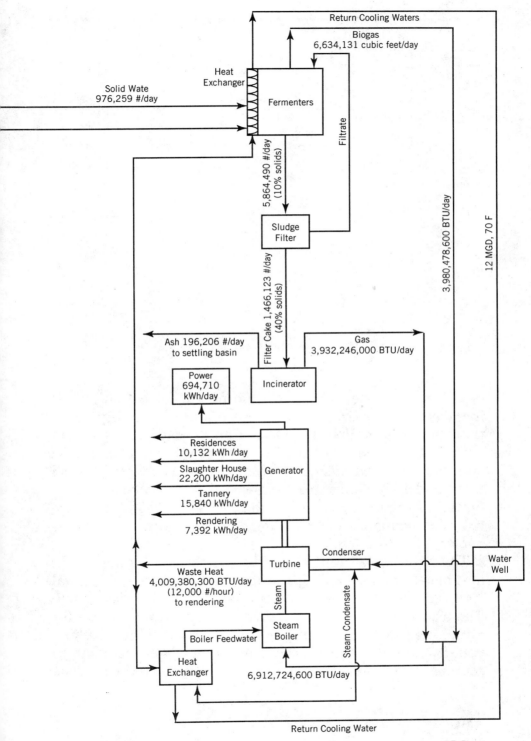

Return Cooling Waters

Biogas
6,634,131 cubic feet/day

Solid Wate
976,259 #/day

Heat
Exchanger

Fermenters

Filtrate

5,864,490 #/day
(10% solids)

3,980,478,600 BTU/day

12 MGD, 70 F

Sludge
Filter

Filter Cake 1,466,123 #/day
(40% solids)

Ash 196,206 #/day
to settling basin

Gas
3,932,246,000 BTU/day

Power
694,710
kWh/day

Incinerator

Residences
10,132 kWh /day

Slaughter House
22,200 kWh/day

Generator

Tannery
15,840 kWh/day

Rendering
7,392 kWh/day

Condenser

Turbine

Water
Well

Waste Heat
4,009,380,300 BTU/day
(12,000 #/hour)
to rendering

Steam

Steam Condensate

Boiler Feedwater

Steam
Boiler

Heat
Exchanger

6,912,724,600 BTU/day

Return Cooling Water

129

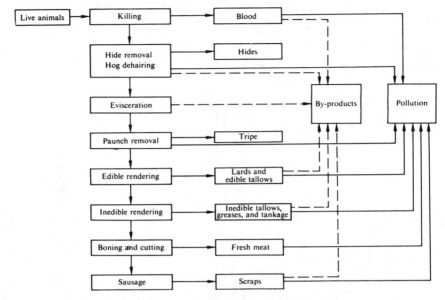

Figure 5.5. Fundamental Processes in the Meat Packing Industry

complex this step can be eliminated, since shipment involves only taking them to an adjacent tannery. Furthermore, no delay is anticipated, because production within the tannery is timed to that of the slaughterhouse supplier. Viscera also are removed and, together with head and feet, bones are delivered to the rendering plant (outlet of the tannery production), as shown also in Figure 5.4. The sequence of operations in the slaughterhouse is shown in Figure 5.5.

The rendering plant converts the fleshing, hide stock, and bones from the slaughterhouse and tannery to animal feed, grease or soaps, and glue, as shown in Figure 5.6. The waste blood is a rich source of protein and, hence, for all slaughterhouses but the very small plant (which usually dumps it into the sewer), it is economical to recover it. The blood is then sent to the renderer in the

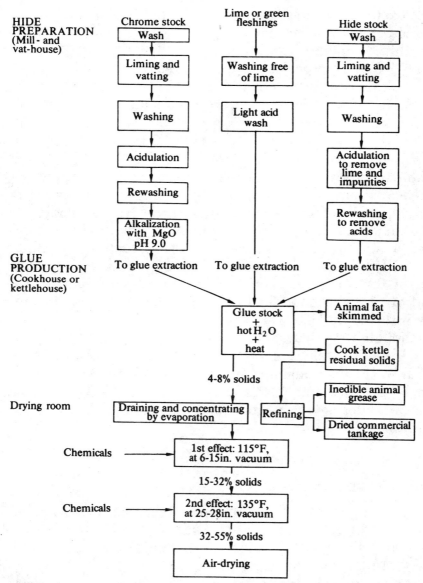

Figure 5.6. Glue Manufacturing Process Flowchart

complex (also without storage or holdup of any type) to produce glue or animal feed (Figure 5.4).

In this complex (Figure 5.4), about 3 million gallons of water, 2.6 million pounds of animal feed, 900 cattle, and 6 tons of chemicals are needed each and every production day. This medium-sized tannery producing 36,000 square feet of leather per day, also provides about 250 tons of meat, 40 tons of tallow, and about 700,000 kilowatts of energy. Methane gas is also produced from all solid waste residues, which is used subsequently for power production. An excess of power within the complex results from this sequence of operations. An alternative to exporting power for sale outside the complex is the production of other valuable intermediate products, such as alcohol derived from the fermenters. A cattle-grazing and feedlot area of 620 acres is required for the 135,000 cattle. In addition, 1,350 tons of feed per day must be supplied from both internal and external sources.

Although a validated analysis of this system has not been made, it appears at least self-sustaining and will probably show a considerable net profit. If the complex is able to produce this profit and protect the environment from any degradation, its major goals will have been achieved.

SUGARCANE COMPLEXES

The Cane Sugar Industry

The cane sugar manufacturing industry is essential to the production of many varieties of foods. In the United States there are about 6,400 sugarcane plantations, 94 sugar mills, and 24 sugar refineries, mostly located in Florida, Louisiana, and Hawaii. Most of the 3 million tons of cane sugar produced each year comes from Florida and Hawaii. Sugar is used in cakes, ice cream, candy, and soft drinks, as well as in other foods and beverages.

Because of the recent dietary recommendations, alternative sweeteners have entered the market. Competition resulting from the lower price of other sweeteners has caused a reduction in refined sugar prices, even though there is a deficit in sugar produced in the United States. Florida, the largest sugar-producing state in the nation, grows about one-fifth of all U.S. sugarcane. It is imperative to the Florida mills, as well as sugar refineries elsewhere, that production costs be kept to a minimum to maintain the health of the industry.

The Sugar Manufacturing Process

In the manufacture of sugar, the sugarcane stalks are chopped into small pieces by rotary knives and the cane juice is extracted from these pieces by crushing them through one or more roller mills. The solid residual material from this operation, consisting of fibrous residue of the cane sugar stalks, is termed *bagasse* and is a solid waste of the cane sugar industry. After the juice is extracted from the stalks, it goes to the boiler room, where lime is added to precipitate insoluble sugars. The precipitate, in the form of a thick slurry, is vacuum filtered to produce a filter cake often termed *cachaza*, which constitutes the second type of solid waste from sugarcane manufacturing operations. Then the clarified juice is thickened in evaporators, and the resulting syrup containing sugar and molasses is boiled in vacuum pans to form raw sugar crystals. The sugar crystals are separated from the molasses by centrifugation, and the molasses is sometimes further evaporated to recover more sugar. The final products are coarse, crystalline brown raw sugar and molasses. The raw sugar is transported for further processing in sugar refineries to produce the various forms of white refined sugar. The bulk of the molasses is used for the production of various types of fermentation products; a

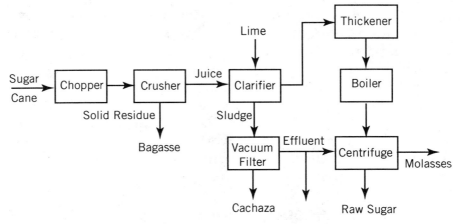

Figure 5.7. Raw Sugar Manufacture—Flow Diagram

small portion is used for animal feed. A schematic diagram of a sugar mill operation is shown in Figure 5.7.

Solid Waste Problem

The two forms of solid wastes generated in the manufacture of cane sugar are, as mentioned earlier, bagasse and cachaza. Every 1,000 tons of processed sugarcane generates about 270 tons of bagasse and 34 tons of cachaza.

The sugar industry is faced with the problem of proper and economical disposal of large quantities of these wastes. The most common disposal method for bagasse has involved burning as much as possible in boilers operated at sugar mills. However, burning bagasse presents problems of its own. It is not a particularly clean fuel, and mills require installation and maintenance of stack scrubbers to clean the emissions. Moreover, the use of bagasse as boiler fuel is impaired by its high degree of moisture (45% to 60%). In addition, its bulkiness requires the construction of special furnaces for efficient operation.

The other type of solid waste generated, cachaza, is generally slurried for disposal by lagooning or disposed of in a landfill, resulting in land and water pollution. Even if a large portion (usually 70%) of the bagasse generated is burned directly in boilers, a considerable amount of it (30%) remains to be disposed of with the entire quantity of cachaza.

Considering the high cellulose content of bagasse and the organic matter in cachaza, these are potential renewable sources of biomass for biochemical conversion to methane by anaerobic fermentation. In addition, the residual digested sludge can have beneficial uses as fertilizer/soil conditioner.

The Environmentally Balanced Industrial Complex Solution

Anaerobic digestion of a 2.4-to-1 mixture of bagasse to cachaza was demonstrated in 1984 to be effective in producing methane gas and reducing organic solids.[17] Despite this development, residual wastes remain to be considered.

An evaluation of the sugarcane refinery based on products and wastes after digestion suggested that a "closed loop" complex would result in the discharge of little or no final residual wastes. Figure 5.8 presents a schematic diagram of a sugarcane refinery-based Environmentally Balanced Industrial Complex. For purposes of this evaluation, the estimated mass balances are based on the refining of 1,000 tons of sugarcane, resulting in the generation of about 270 tons of bagasse and 34 tons of cachaza.[18] In many mills, these wastes are discharged to the environment with a variety of adverse impacts.

---

[17]N. L. Nemerow, "Environmentally Balanced Industrial Complexes" *The Biosphere: Problems and Solutions* (Elsevier Science) (London: B.V. Amstermdam, 1984), 461–470.
[18]Ibid., pp. 463–465.

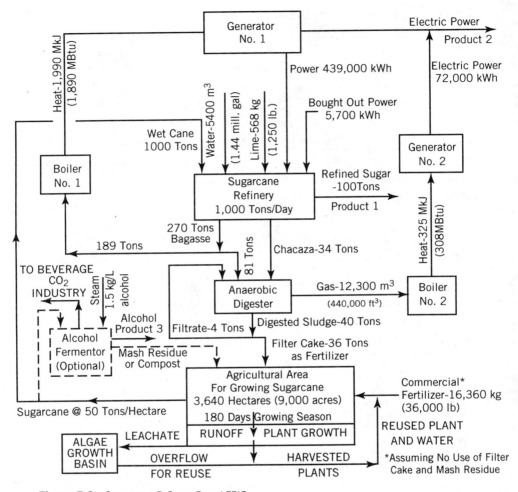

Figure 5.8. Sugarcane Refinery-Based EBIC

In the refinery proposed in this complex, 189 tons of bagasse are burned to supplement the steam required in production. Consequently, 81 tons of wet bagasse and 34 tons of cachaza remain as waste products. In addition, during the growing and harvesting of sugarcane, fertilizer and pesticide residues are washed off the growing area (as nonpoint pollution runoff) and into a runoff collection basin sloped to drain excess water into an algae growth basin. Plant growth (from the runoff basin) and algae (from the algae growth basin) are reused in the sugarcane growing area for fertilizer, along with excess growth basin overflow water to enhance cane growth (see Figure 5.8).

Anaerobic digestion of the bagasse-cachaza mixture generates about 12,300 cubic meters (440,000 cu/ft) of gas (about 70% methane), 36 tons of filter cake, and a filtrate that is recirculated to the digester to enhance the digestion process. The gas that is burned in boiler No.2 produces 325,000,000 kJ (308,000,000 Btus) of steam for energy production. This steam can be bled off to supply the refinery or used to generate 72,215 kwh of electricity, which can be used in the refinery or sold as a product outside the complex. As an illustration of this power use, the Okeelanta Co-generation Project will burn enough sugarcane bagasse and wood waste to power 46,000 homes when completed in early 1996. It will be the nation's largest biomass cogeneration project in the United States, according to J. McNair.[19] The plant will burn only renewable resources. Two-thirds of the fuel will come from bagasse, the rest from scrapwood bought from builders and demolition firms in the area.

Thirty-six tons of filter cake are used as fertilizer to replace a portion of the commercial fertilizer needed in the fields to grow the cane. Since the filter cake may not contain all the requisite types and quantity of nutrients of a commercial fertilizer on a unit-weight basis, an appropriate amount of commercial fertilizer will be mixed with it to meet the necessary growth requirements, depending on the nature of the soil. Approximately 3,640 hectares (9,000 acres) of land are required for harvesting 180,000 tons of

---

[19]J. McNair, "Sugar Cane Will Fuel Okeelanta Power Plant," *The Miami Herald*, 17 February 1994, 3C.

cane as feedstock for the refinery at a rate of 1,000 tons per day. The mill operates only during the annual sugar-cane growing and harvesting season of 180 days. Since the mill runs continuously, 24 hours a day, seven days a week, during the cane-growing season, it is customary to operate and check out the equipment thoroughly during the rest of the year to prepare for the next year's operation. The number of hectares harvested per day will vary, depending on the daily requirements of the refinery. An average of 50 tons of wet cane is obtained per hectare planted and harvested (20 tons per acre). Depending up the market for refined sugar, some cane can be used to feed the alcohol fermenter, which can produce one liter of alcohol for very four pounds of cane. For each liter of alcohol produced, 1.5 kg (3.3 pounds) of steam are required. The steam can be supplied by either boiler No. 1 or No. 2. The mash residue from the alcohol fermenter can also be used as supplemental fertilizer for the sugarcane fields, further decreasing the amount of commercial fertilizer needing to be purchased from outside the complex.

Another potential configuration of the sugarcane complex is shown in Figure 5.8A. In this complex the nonpoint drainage water wastes from the sugarcane growing field are collected in special drainage canals in a sump at the lowest elevation. From the sump, the growing-field wastes are pumped back to the cane-growing area to be used as irrigant and nutrient material. Excess runoff wastes are fed to the chemical treatment center where lime is fed and the wastes are coagulated and finally settled. Both the supernatant and the lime sludge from the coagulation treatment are returned to the cane-growing field as additional cane nutrients to augment or supplement commercial fertilizer. When the fields are saturated with water, the clear supernatant is returned to the sugar mill for reuse in processing the cane. Therefore, no fertilizer

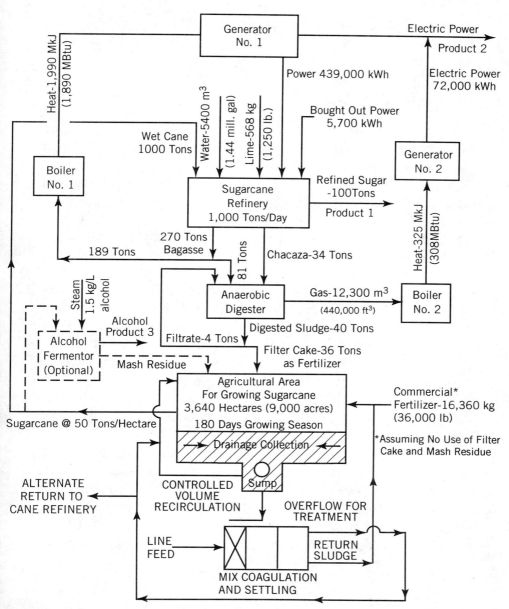

Figure 5.8A. Sugarcane Refinery-Based EBIC

(nitrogen and phosphorus) or insecticide wastes leach from the growing area into the surrounding water environment.

TEXTILE COMPLEXES

The textile industry represents one of the most competitive areas of worldwide production. Each plant tries to reduce its cost to a minimum in order to compete with other similar plants wherever in the world they are located. One way a mill can do this is by increasing production either in its own plant or by merging with another mill. Lower unit costs generally result from increased production in accordance with accepted economic principles.

Many textile mills are small in size, vital to the regional economy, and find it very difficult to compete with each other and with larger mills. These mills desperately need to find ways to lower production costs. Many of them are also located on small watercourses where their wastes take an unusually high toll on the environment. Governmental pressure is being applied to avoid and avert such pollution, yet conventional waste treatment will increase production costs.

Considering the relatively small size of most textile mills and the environmental pollution they cause, a more economically feasible solution must be found. Further, the supply of fresh, clean raw water at low prices is dwindling. The ultimate survival of textile mills—especially small ones—depends on solving both the economic and the environmental problems.

Textile Mill Operations

Textile mill operations consist of weaving, dyeing, printing, and finishing. Among the raw materials, mainly cotton, wool, and synthetic fibers, cotton is of primary concern to environmentalists. Many processes involve several steps, each contributing a particular type of waste: sizing of fibers, curing (alkaline cooking at elevated temperatures), desizing the woven cloth, bleaching, mercerizing, dyeing, and printing. Textile wastes are generally colored, highly alkaline, high in BOD and suspended solids, and high in temperature. The sequences of processes for cotton and wool are shown in Figures 5.9 and 5.10 respectively. Wastes from synthetic fiber manufacture resemble chemical manufacturing wastes, and their treatment depends on the chemical process employed in the fiber manufacture. Cellulose and noncellulose synthetic fiber processes are also shown schematically along with significant pollutants in Figures 5.11 and 5.12. These processes are described in greater detail elsewhere.[20]

Two production costs that influence the selection of an EBIC system for textile mills are the costs of (1) raw process water and (2) wastewater treatment. First, although the cost of process water generally represents a minor portion of total manufacturing costs, it is significant because it is becoming an increasing percentage. Process water is also becoming a scarcer raw material. In general, municipal water utilities charge from \$0.50 to \$1.50 per 100 ft^3 (750 gallons) of water. For a typical small textile finishing mill producing woven fabric in a series of complex processes that use 600,000 gallons per day, the daily cost would vary from \$264 to \$858. Even these charges may be misleading, because they occurred only where this amount of water was available for sale.

Second, conventional wastewater treatment in small textile finishing mills has been either (a) separate treatment and reuse of dye wastes only or (b) complete treat-

---

[20]N. L. Nemerow, *Industrial Water Pollution*, 311; and N.L. Nemerow and A. Dasgupta, *Industrial and Hazardous Wates*, 313–322.

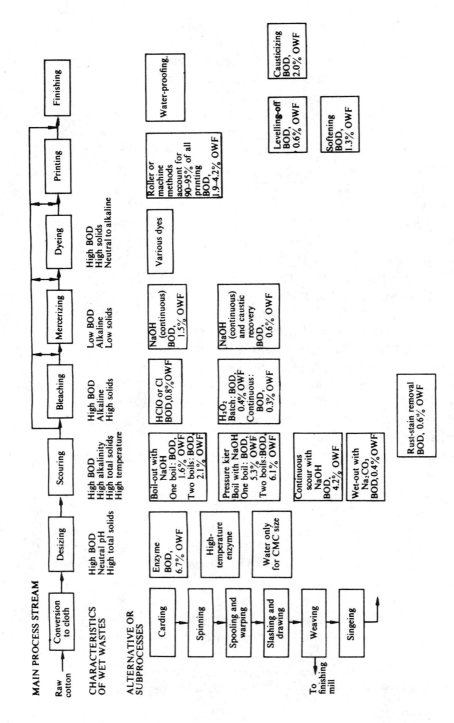

Figure 5.9. Cotton Textile Finishing Process Flowchart

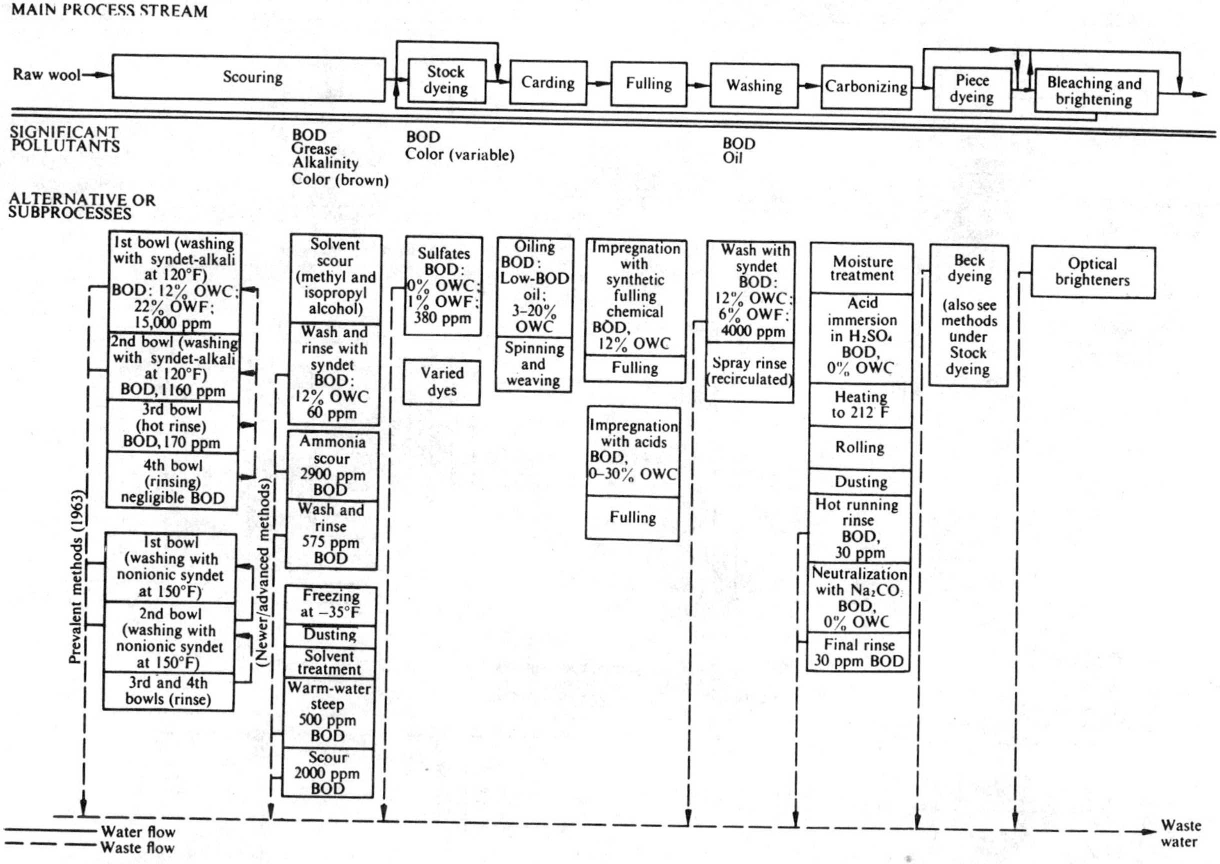

Figure 5.10. Wool Textile Production Process Flowchart

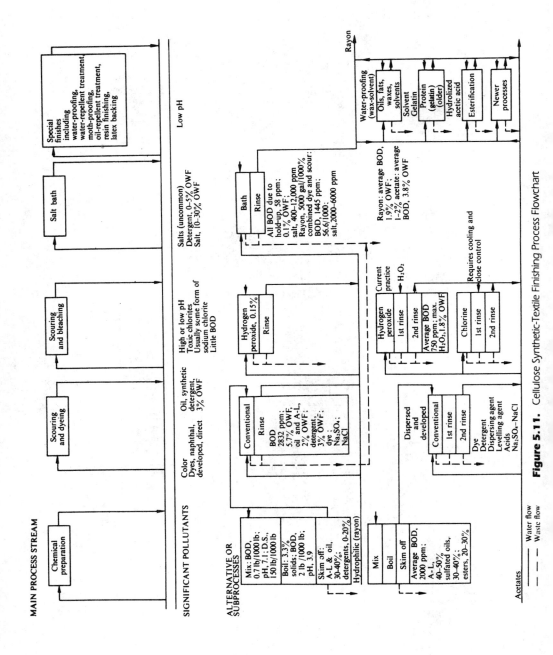

Figure 5.11. Cellulose Synthetic-Textile Finishing Process Flowchart

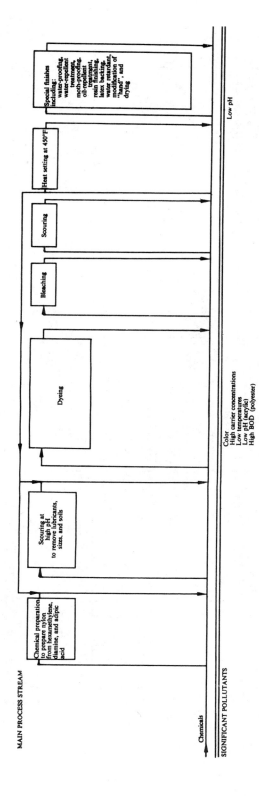

Figure 5.12. Non-Cellulose Synthetic-Textile Finishing Process Flowchart

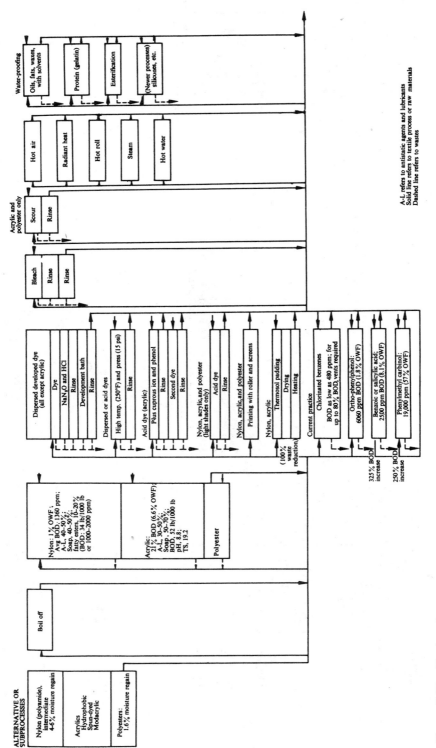

Figure 5.12. (Continued)

ment of the total finishing mill waste.[21] The first has been accomplished mainly by hyperfiltration and/or dye-bath reconstitution, and the second has been done mainly by chemical coagulation and/or biological aeration.

Both methods have produced certain amounts of reusable water. However, economic considerations as well as government environmental regulations play major roles in the decision to produce reusable wastewater. The cost to the small mill of producing acceptable-quality reusable wastewater will need to be further defined and reduced to a minimum before reuse becomes standard practice regardless of receiving-water quality degradation.

Costs of Conventional Wastewater Treatment

The capital and operating costs of small textile mill wastewater treatment depends largely on the type and extensiveness of the treatment used. Capital costs range from as low as $31,500 for simple dye bath reconstitution to as high as $303,000 to $982,000 for chemical coagulation, filtration, and activated sludge treatment. Operational costs for similar treatments range from $40,000 to $328,000 per year.

A typical small mill produces about 25,000 pounds per day, with an average capital cost of $500,000 for complete treatment and an annual operating cost of $150,000. The result is capital costs of $20 per pound of production per day and an operating cost of $.02 per pound of production per day (assuming 6 days per week and 50 weeks per year of production). These are only approximate costs for presumed average small mills; the range of true costs may vary considerably from these approximations. However, it is apparent that both capital and operating costs of these small mills represent a very significant expenditure.

---

[21]N. L. Nemerow and A. Dasgupta, *Industrial and Hazardous Wastes.*

Minimization of these production costs by including treatment costs as a negative benefit of wastewater reuse would constitute a real boon to the small mill.

Alternate Solutions to the Dilemma

There are two potential methods of reducing the waste treatment costs of a small textile plant and, at the same time, producing reusable wastewater to replace or replenish the mill's costly water supply: (1) industrial complexing and (2) chemical coagulation (only industrial complexing will be considered in more detail in this chapter). Other methods reported in the literature may reduce waste treatment costs or produce a partial supply of raw water, but will not accomplish both objectives. For example, dispersed growth aeration, as we suggested in 1985,[22] will treat the wastewater at reduced costs, but will not, by itself, produce acceptable reusable water. In 1976 Brandon and Porter[23] hyperfiltered dye wastes through membranes to produce both recyclable water and dyes, but failed to treat a sufficient portion of the plant's total waste at a lowered cost to result in satisfactory overall waste treatment. To be cost-effective for the small textile manufacturer, the solution to its problems must satisfy both environmental and production concerns.

Industrial Complexing

Water-consuming and waste-producing textile finishing mills are ideally suited for industrial complexes. Although the wastes of this industry may pollute our fragile environment, they may be amenable to reuse by close associa-

---

[22]N. L. Nemerow and A. Dasgupta, "Zero Pollution for Textile Waste," *Proceedings of the 7th Conference on Alternative Energy Sources*, vol. 6, Hemisphere Publishing, 1987, p. 499.

[23]C. A. Brandon and J. J. Porter, *Hyperfiltration for Renovation of Textile Finishing Plant Wastewater*, EPA-600-2-76-060, (Washington, D.C.: Environmental Protection Agency, 1976).

tion with satellite industrial plants that are able to use them and, in turn, produce raw materials for others within the complex.

An ideal, illustrative EBIC for the small textile finishing mill is shown in Figure 5.13. This complex contains five manufacturing plants producing 12,000 pounds of woven fabric for sale outside the complex and 13,200 pounds of cotton, 14,640 pounds of greigh goods, 1,952 pounds of NaOH, and 120 pounds of dyes for reuse within the complex.

In addition, all sewages and wastewaters are reused without treatment within the complex. Of notable interest and importance is the reuse of 86,434 gallons of untreated finishing mill desize waste, which contains 732 pounds of BOD.

A raw material balance justification is presented in Table 5.2, which compares the raw material quantities and costs for the five separate plants manufacturing at distant locations and manufacturing within the EBIC. These data, although obtained and referenced only from authentic literature, show the cost advantage of this textile EBIC.

In addition, environmental costs would have to be considered. Within the industrial complex we are presuming no external environmental costs are needed. For separate plants operating at distant locations from each other, environmental costs would include both domestic and industrial waste treatment charges, as well as any measurable adverse environmental impact costs for the residual effluent wastes.

In Table 5.2 it is apparent that the cost savings of the industrial complex, from a material balance alone, is $6,726 minus $6,093.75 or $632.25 per day, which represents a savings of $52.69 per 1,000 pounds of finished cotton fabric.

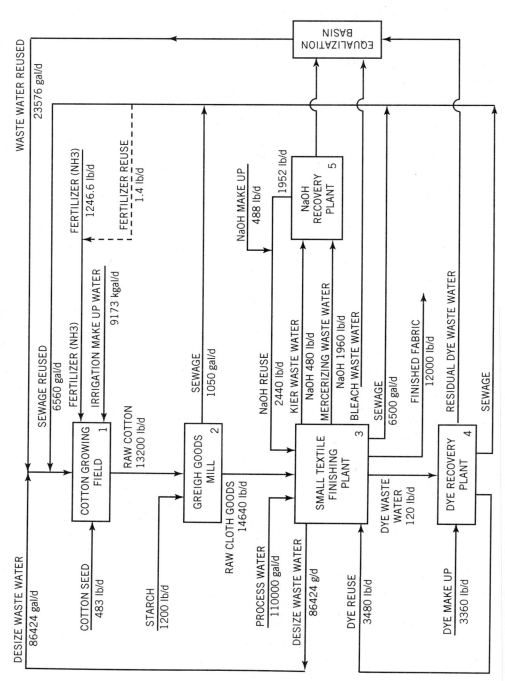

Figure 5.13. Diagram of the Integrated Five-Plant Industrial Complex

Table 5.2 Raw Material Balance—As Part of Total Production Cost

Raw Materials Needed / Amount Needed	Plant Type / Amount Needed	Mass Balance of Small Textile Complex [a] Raw Material				Cost/Unit of (1982-1983)
		By the Complex Quantities	Cost($)	When Plants Are Separate Quantities	Cost($)	
	Agricultural growing field					
Irrigation water		9.173mgd	4,036.00	9.29mgd	4,087.00	$0.44–1.43/1000gal
Cotton seed		483#/day	48.30	483#/day	48.30	0.10#
Fertilizer		1,248.6#/day	81.16	1,250#/day	81.25	0.065# (1987)
	Greigh goods Manufacturing					
Starch sizing		1200#/day	180.00	1200#/day	180.00	$0.15
	Textile mill Finishing					
NaOH for fabrics		488#/day	131.80	2,440#/day	658.80	$0.27#
Dyes for fabrics		3,360#/day	1,512.00	3,480#/day	1566.00	0.45#
Process water		0.11 mgd	104.50	0.11 mgd	104.50	0.44–1.43/1000gal
Total Material Cost			$6,093.75/day		$6,726/day	

[a]Assume Typical Small Textile Finishing Plant Producing 12,000 #/day of woven finished cloth product.

From the preliminary study the total environmental savings appear to be greater than $1,248 per day, which represents the average costs of treatment for separate greigh goods and finishing mill wastes. To both savings we must add the savings from transportation of raw cotton sized woven goods. Presuming the transportation cost is $0.026 per pound of cotton transported, we can estimate the additional cost of transportation as $343.2 per day.[24]

The cotton growing field is included in Figure 5.13 as the very first portion of the textile complex. This necessitates that the fertilizer to grow the cotton be a vital component of the complex. In fact, in the mass balance shown in Figure 5.13, 1,248.6 pounds of ammonia fertilizer are required. The portion of this material that can be replaced by nitrogen (as well as phosphorus and other nutrients) contained in the reused equalization basin effluent basin is unknown at this time. However, the added nutrients should prove significant and helpful. This will be important to make this complex as self-sustaining as possible. "Sustainable agriculture" is proposed by Reganold, Papendick, and Farr as a solution to the conservation of natural resources, as well as economically wise. For a farm to be sustainable, these authors maintain, adequate food of top quality must be provided and its resources must be environmentally safe as well as profitable. They proclaim that this type of farming relies greatly on beneficial natural processes and renewable resources coming from the farm itself. They also say that in this case the costs of production are lower than those of nearby conventional farms. They report that the board of agriculture of the National Research concluded that "wider adoption of proven alternative systems would result in even greater economic benefits to farmers and environmental gains to

---

[24]N. L. Nemerow and A. Dasgupta, "Zero Pollution for Textile Waste," pp. 499–508.

the nation." Presumably such benefits and gains would result from industrial complexes that include agricultural activities to reduce environmental impacts.[25]

FERTILIZER–CEMENT COMPLEXES

The fertilizer industry, which wastes calcium and sulfate while producing phosphoric acid, and the cement industry, which utilizes calcium as a raw material while producing a calcium oxide cement, are ideally compatible plants for an industrial complex. To understand the needs and goals of these two plants, we must delve into their separate production requirements.

The phosphate fertilizer industry in producing phosphoric acid from phosphate rock deposits leaves a by-product of about five tons of phosphogypsum ($CaSO_4 \cdot 2H_2O$) for each ton of phosphoric acid produced.[26, 27] Worldwide yearly production of waste phosphogypsum (PG) is estimated at 150×10^6 tons, of which very little is reused.[28] Many fertilizer plants worldwide have been denied permits, and existing plants not allowed expansion, because of the vast amount of land required for waste sludge storage. In addition, runoff or leachate from the phosphogypsum piles can pollute surrounding water resources. In central Florida, which accounts for about 75% of U.S. needs and 33% of the world's supply, stockpiles will amount to more than 1 billion tons by the year 2000.[29, 30] When one couples the phosphogypsum buildup

---

[25]J. P. Reganold, R. I. Papendick, and J. F. Farr, "Sustainable Agriculture," *Scientific American* (June 1990): 112.

[26]W. F. Chang and M. I. Mantell, *Engineering Properties and Construction Applications of Phosphogypsum* (Coral Gables, Fla.: University of Miami Press, 1990), Preface page.

[27]Nemerow and Dasgupta, *Industrial Hazardous Wastes*, p. 223.

[28]Chang and Mantell, Preface page.

[29]Ibid., Preface page.

[30]Nemerow and Dasgupta, *Industrial Hazardous Wastes*, p. 223.

with the need for cement and associated building products for the worldwide construction industry, the proposed industrial complex appears to be a "natural."

After the rock is mined, slurried, and separated from clay and sand by screening and flotation, it is used to produce wet-process phosphoric acid. The rock is digested by the action of sulfuric acid (a required raw material) and produces a slurry of contaminated gypsum ($CaSO_4 \bullet 2H_2O$) and phosphoric acid (H_3PO_4). The slurry discharges over a rotating disc filter whereby the calcium sulfate and the acid are separated. The waste gypsum is pumped to holding ponds, where it represents a major disposal problem for the fertilizer industry. The acid is concentrated and reacted with ammonia to produce ammonium phosphate (($(NH_4)_3PO_4$) fertilizers. Wet-process acid is also reacted with the phosphate rock to produce triple concentrated superphosphate (TSP) fertilizers.

About 63% to 65% of wet-process acid is used to make ammonium phosphate, 4/5 as diammonium phosphate (DAP) and 1/5 as monoammonium phosphate (MAP). Another 10% to 15% of the wet phosphoric acid is used to make TSP. Producers prefer DAP because it is easy to transport and handle, relatively concentrated, stable, and not too expensive to make. Farmers also like DAP because it can be uniformly applied and contains both nitrogen and phosphorus, two essential fertilizer ingredients.

Any recovery and reuse system for phosphogypsum will release reclaimable land for more productive uses. Further economic benefits will be derived by elimination of adverse environmental consequences of leachates and runoff from the gypsum heaps. Leachates carry phosphate and mineral nutrients that may pollute drinking waters and cause algal blooms (red tide) in recreational waters. Direct reuse of phosphogypsum for external purposes presents the dilemma of incorporating these impurities, along with radioactivity, into building or road products. It is quite likely that the direct use of phosphogypsum within a closed industrial complex to make cement would eliminate all of these problems and lower production costs of both plants. A phosphate fertilizer complex also must dissipate large amounts of waste heat. The recovery and utilization of this lost energy (waste heat) within the complex will also result in additional economic benefits.

To understand and establish the exact characteristics of the waste phosphogypsum that we intend to reuse as a raw material for cement, see the analysis given in Table 5.3.

Chang and Mantell report that free acids, phosphates, and fluorides interfere with the direct reuse of waste gypsum to make gypsum board (the largest reuse of PG). These impurities affect the setting time and strength. Only about 19% of the waste gypsum is reused as an additive to cement. As reported by these authors, adding 3–5% gypsum to cement retards the setting time of the cement, counteracts shrinkage, and provides a quicker development of initial strength, higher long-term strength, and greater resistance to sulfate etching.[31] Impurities in waste gypsum (phosphates, fluorides, and organic matter) must be considered since they affect the cement quality.

G. Erlenstadt also reports that the 20% to 30% water content of waste phosphogypsum makes its transport (to a distant cement plant), as well as its storage and proportioning in cement mills, extremely difficult.[32]

D. A. Clur claims that the Fedmis (South Africa) fertilizer plant disposes of about 25% of its PG production as soil conditioner, cement clinker, and cement retarder.

---

[31]Chang and Mantell, p. 6.

[32]G. Erlenstadt, "Upgrading of Phosphogypsum for the Construction Industry," *Proceedings of International Symposium on Phosphogypsum*, Florida Institute of Phosphate Research, University of Miami, Barlow, FL (1980), 284–293.

Table 5.3 Typical Analysis of Phosphogypsum

	[a]Dihydrate (%)	Hemihydrate (%)	Hemidihydrate (%)
CaO	32.5	36.9	32.2
SO_3	44.0	50.3	46.5
P_2O_5	0.65	1.5	0.25
F	1.20	0.8	0.5
SiO_2	0.5	0.7	0.5
Fe_2O_3	0.1	0.1	0.05
Al_2O_3	0.1	0.3	0.3
H_2O of crystallization	~19.0	~9.0	~20.0

[a]Most commonly used worlwide and in the United States.
Source: W.F. Chang and M. I. Mantell, *Engineering Properties and Construction Applications of Phosphogypsum* (Coral Gables, Fla.: University of Miami Press, 1990), 5 and 6.

"The quality of the cement compares favorably with that of local limestone-based cements, and is used in all classes of building construction and civil engineering." Clur also reports that "the technical problems of producing good quality cement from PG [having] largely been solved, the future of the process would seem to depend mainly on economical and environmental factors."[33]

The cement industry uses calcines, a mixture of calcareous and argillaceous materials in the proper ratio. For example, we have included in Table 5.4 the raw materials actually consumed in the United States to make Portland cement (the most commonly accepted type) in 1972.

The possibility of using waste phosphogypsum as a raw material replacing limestone as the source of calcareous matter in manufacturing cement is of utmost importance to the success of the proposed complex. Therefore, we present at this point some of the processes and materials conventionally used in making cement.

---

[33]D. A. Clur, "Fedmis Sulfuric Acid/Cement from Phosphogypsum," Second International Symposium on Phosphogypsum, Florida Institute of Phosphate Research, University of Miami, Barlow, FL (1986), 141.

Table 5.4   Raw Materials Consumed for Portland Cement in the United States (1972)

	(1,000s tons)	(%)
Cement rock	23,799	23.8
Limestone	90,003	90.0
Marl	2,080	2.1
Clay and shale	12,158	12.2
Blast furnace slag	759	0.8
Gypsum	4,094	4.1
Sand and sandstone	2,774	2.8
Iron materials	839	0.8
Miscellaneous	414	0.4
Total		100.0

Source: R. N. Shreve and J. A. Brink, Jr., *Chemical Process Industries*, 4th ed. (New York, Mc Graw-Hill, 1977), 159.

Cement making involves storage and mixing of the materials indicated in Table 5.4, drying, grinding and crushing, calcining, clinker storage, adding finishing materials, ball milling of the clinker and additives, and packaging for sale. Dry processing is usually practiced but wet processing is also used. A schematic of both processes is shown in Figure 5.14, and a process incorporating modern technology in Figure 5.14A.

Shreve and Brink[34] report that for each 376 barrels of finished cement produced by the dry process 1,120,000 Btu's of fuel are required, as well as 24.1 kwh of electricity, 30 gallons of water, and 0.17 hours of direct labor. Raw materials required for the 376 barrels are 498 pounds of limestone, 124 pounds of shale, and 16 pounds of gypsum. An average of 4 pounds of $CaCO_3$ for each pound of clay has also been reported by various researchers.

One potential configuration for a fertilizer–cement plant complex designed to eliminate environmental adversity is shown in Figure 5.15. In this complex, waste phosphogyp-

[34]R. N. Shreve and J.A. Brink, Jr., *Chemical Process Industries*, 4th ed. (New York: McGraw-Hill, 1977), 162.

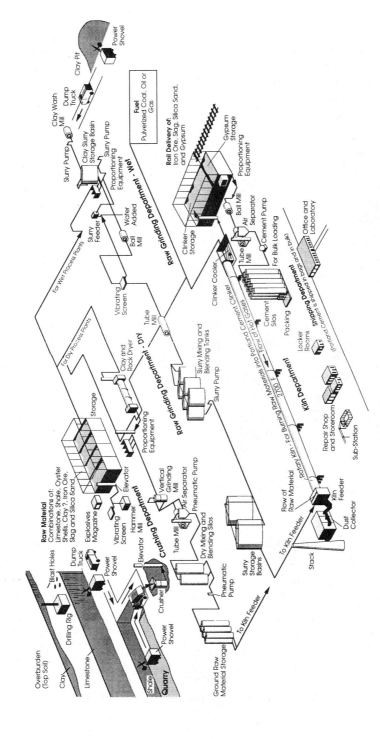

Figure 5.14. Isometric Flowchart for the Manufacture of Portland Cement

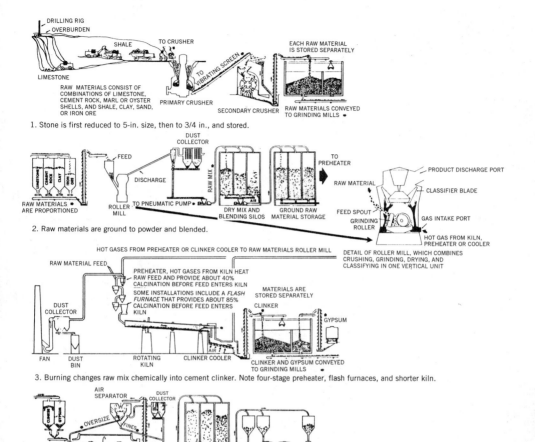

1. Stone is first reduced to 5-in. size, then to 3/4 in., and stored.

2. Raw materials are ground to powder and blended.

3. Burning changes raw mix chemically into cement clinker. Note four-stage preheater, flash furnaces, and shorter kiln.

4. Clinker with gypsum is ground into portland cement and shipped.

Figure 5.14A. New Technology in Dry-Process Cement Manufacture

sum is used directly by the cement plant while the latter's waste of SO_2 gas and waste heat from the calciner are reused as well. The sulfuric acid produced by scrubbing the SO_2 is also reused by the fertilizer plant through dissolving the phosphatic rock raw material. The hot water (waste heat is reused by the composter to speed the

160

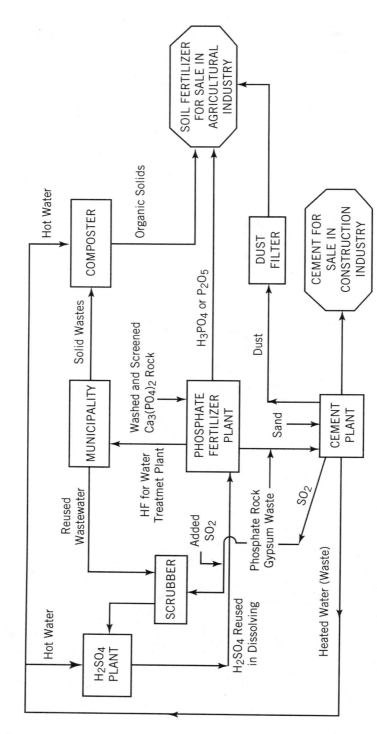

Figure 5.15. Fertilizer-Cement Plant Complex

reaction in conversion of municipal solid wastes to compost. Dust from the cement plant is also incorporated in the compost fertilizer product for the agriculture industry. Even the small amount of hydrofluoric acid condensed from the fertilizer plant stack gas is proposed to be reused by the municipality to fluoridate its water supply. After conventional municipal sewage treatment, the effluent will also be used as scrubber water.

Research is currently under way to ascertain the feasibility of this and other schemes for a balanced complex, as well as to determine the mass balance for the various operating plants.

Portland cement, which requires about 6.3 million Btu's of energy to produce each ton,[35] exacts the following physical and chemical characteristics for acceptable marketing.[36]

Physical Properties Specified Limit

Minimum compressive strength
3 day 1,800 psi (12.8 M Pa)
7 day 2,600 psi (18.5 M Pa)
28 day 3,800 psi (27.0 M Pa)
Time of set (Vicat)
Initial (minimum) 0.75 hr
Final (maximum) 8.0 hr
False set, final penetration
(Minimum) 50%
Autoclave expansion (maximum) 0.80%

---

[35]R. M. Lea, *The Chemistry of Cement and Concrete* (New York: Chemical Publishing Co., 1971).

[36]"Standard Specifications for Portland Cement," *1975 Annual Book of ASTM Standards*, Part 13 (Philadelphia: American Society for Testing and Materials, 1975).

Chemical Properties Maximum Allowable Weight %

Alkalies 0.6

Chloride 2.0

Lead oxide (PbO) 0.001

Magnesium Oxide (MgO_2) 5

Sulfate (SO_4) 3

It is important for our purpose to understand the limitations posed by three potential chemical contaminants of cement in making concrete. First, excessive sulfate levels may cause "efflorescence" when used in concrete. Efflorescence is the loss of water of hydration from a soluble salt that has migrated to the surface of the concrete, which results in a white coating on this surface. Sulfur compounds in the raw material (phosphogypsum after calcining) limit the cement plant's ability to add $CaSO_4$ to control the setting time of the cement. Second, alkalies react with certain aggregates (especially silicious ones) to form an alkali silicate gel that causes expansive forces in the concrete, leading to cracking and failure of concrete sections. Third, chloride levels are particularly important when used in conjunction with cement or concrete. Chloride tends to cause corrosion of steel, and a chloride level of 2% or less in cement in often specified.

It appears feasible to locate, build, and operate a two-industry complex consisting of a phosphate fertilizer plant and a cement plant. This could lower the production costs at both plants and eliminate all adverse environmental impacts at the same time.

Current research is aimed at determining (1) the optimum size for each manufacturing plant included within the complex, (2) the suitability of the by-products (wastes) for recovery and reuse as raw materials for ancillary adjacent plants within the complex (compatibility of plants),

(3) the validity of the total waste elimination from the two plants within the complex, and (4) the actual cost of production of the prime goods when manufactured at distinctly separate plants (See Figures 5.16 and 5.17) as compared with the same when manufactured within the complex, as shown in Figure 5.15. When computing the economic gain by using this "complex" principle, one must include the economic cost of environmental damage caused by wastes of all plants involved as part of the production costs.

In understanding this complex, the reader should be aware that solid waste is only a misplaced raw material. Some particular industry can use the misplaced raw material to make a valuable product. The problem arises in the matching of the waste producer with the raw material user. Once a producer and user are located, three questions must be resolved:

1. What is the physical distance between the two?
2. What are the economics of joining the two plants?
3. Are the waste materials from each plant useable as raw materials for the other plant?

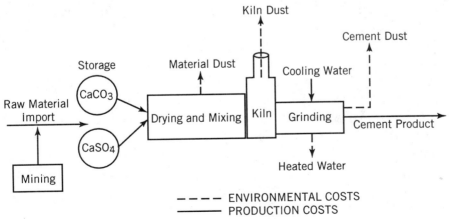

Figure 5.16. Isolated Cement Plant

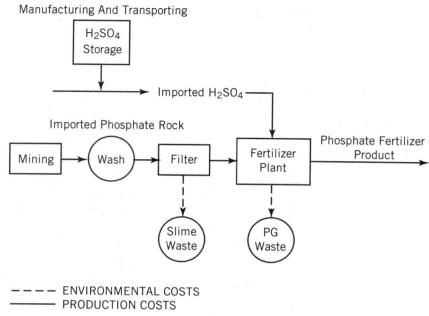

Figure 5.17. Isolated Phosphate Fertilizer Plant

Distance

Industrial plants, up to now, have generally considered locating where factors such as raw materials, labor, market, taxes, utilities, community cooperation, transportation access, and so forth, are favorable. These factors may place our two suitors many miles apart, since their reaction to each of these factors will often differ from industry to industry. We must prove to each of them that it is in their best interests to locate adjacent to each other. In the final analysis, it must be less expensive for the two compatible plants to locate close to each other in the same complex.

Economics

Industry wants and deserves facts that prove it is less expensive to manufacture two or more plants' products in the same complex than at locations distant from one another. The only way to provide this verification is to make a prototype analysis, generate real data, and feed these data into an analytical model to yield final production costs. Of key importance in determining the optimum solution are the real costs of avoiding pollution of air, water, and land. These costs must be quantified and included in the analytical model.

Compatibility

The two (or more) plants included in the EBIC must be compatible. Mainly, this refers to the acceptability of one plant's waste (or wastes) as raw material(s) for the other's production. The waste will usually substitute for some or all of one raw material. This may mean the acceptance by the other plant of somewhat altered manufacturing techniques or slightly different production specifications. Sometimes wastes, as raw material, may require some modification prior to use by the receiving plant.

Compatibility also refers to the cooperative abilities of the plants within the complex. Each plant becomes dependent on the other(s). Factors that affect the production of one plant will affect its wastes, either by volume or character, and these, in turn, will affect the production of the other plant(s). Machine or utility breakdowns, fires and other catastrophes, changed market conditions, temporary labor strife, and so forth, will all affect production schedules and the availability of wastes, which must be considered by the ancillary plant(s) in the complex. These situation also arise when cooperating plants are at some distance from each other, or when such as when the delivery of raw materials is delayed. In fact, it may be easier to ameliorate these reasons within a complex than at plants distant from each other. In any event, all plants must be cooperative, understanding, and flexible enough to operate when unusual situations develop.

The ideal solution to waste problems of all kinds, gaseous, water, and solid, is the ultimate recycling or complete reuse within an industrial complex. This produces zero pollution, affects no environment external to the complex, and results in lowered production costs for the industrial plants, and possibly greater profits. In this book, I describe briefly some groupings of compatible industrial plants, and show how the wastes produced are used by other plants. However, most of these complexes are theoretical at this time and are not engaged in actual production, but indications are that some are getting closer to implementation.

Phosphate fertilizer and cement plants represent typical examples for this research, since both produce valuable products, not only for Florida and India where raw materials are mined, but also for all people worldwide. They also produce liquid, solid, and gaseous wastes. In most cases such plants are located at a considerable distance from each other. These are ideal candidates for an Environmentally Balanced Industrial Complex. This complex can best be visualized by the three figures in this section.

Figure 5.16 depicts a conventional cement plant located in an isolated spot so as not to receive complaints from neighbors about environmental pollution. The cement plant imports mixed limestone, sand, clay, and even some steel mill blast slag and gypsum. Major wastes emanating from typical cement plants are kiln dust, heated kiln cooling water, and hot combustion gases. Some production and environmental costs that result from the use of these raw materials and discharge wastes include the following:

1. Mining limestone
2. Transporting the limestone to the cement plant

3. Storing the limestone at the cement plant
4. Mining, transporting, and storage of any gypsum used
5. Disposal of collected kiln dust
6. Adverse effect of heated cooling water on the receiving watercourse.
7. Adverse effect of CO_2 on the atmosphere

In Figure 5.17, a conventional phosphate fertilizer plant is shown, also located in an isolated spot convenient to phosphate rock mining and away from residences. This plant mines (or imports mined) phosphate rock, which is then slurried and separated from clay and sand by screening and flotation. The rock is then digested by the action of sulfuric acid. This produces H_3PO_4 and a contaminated waste sludge, $CaSO_4 \cdot 2H_2O$ (phosphogypsum). After separation of the two materials, phosphoric acid and phosphogypsum, the waste gypsum is normally pumped to holding ponds where it stands and piles up, representing a major disposal problem. Some production and environmental costs that result from the use of these raw materials and discharge of wastes include the following:

1. Transportation and storage costs of H_2SO_4.
2. Potential of environmental disasters with importation of H_2SO_4.
3. Land use cost for piling up and storing phosphogypsum (PG).
4. Leachate damage to underground water supplies from PG.
5. Surface runoff damage to water resources from PG. Some of the potential water contaminants from stored PG are fluorides, radon radioactivity, phosphate nutrients, hardness, and silt.

All of these mining, transportation, potential environmental damage, and storage costs add up to make the two major products, cement and phosphate fertilizer, too costly. In addition, the fragile environment surrounding these plants is damaged to varying degrees by their productions.

The solution to both the excessive production costs and the environmental damage is to join production of both cement and phosphate fertilizer in one complex. An example of such an complex is shown in Figure 5.15. In this EBIC, no waste leave the complex. All added costs of mining, transportation, damage, and spills during shipping and storage of imported raw materials are omitted. Production costs are reduced and adverse environmental impacts are eliminated.

FOSSIL-FUELED POWER PLANT COMPLEXES

Dilemma of Electric Power Plants

Electric power plants face the problem of producing more electricity at a lower production cost and, at the same time, minimizing damage to the surrounding environment. This is extremely difficult to accomplish because of the problem of obtaining permits for producing nuclear power, the polluting characteristics of both oil and coal fuels, and the untried utilization of the more sophisticated wind, solar, and hydrogen-generated power. Since fossil-fueled power plants are currently cost-effective and generally acceptable to public, it is reasonable to use this form of fuel and attempt to ameliorate or abate the adverse environmental consequences. The challenge is to do this effectively, that is, at a minimum production cost and with little or no adverse environmental impact.

COAL-FIRED POWER PLANTS

Coal-fired power plants generate most of the electricity in the United States. In 1986 the electric utilities produced 2,487.3 billion kilowatt-hours of electricity. In the same year, coal-fired plants generated 1,385.1 billion kilowatt-hours of electricity. This was approximately 56% of the nation's total production for the year.[37]

The United States Department of Energy has been encouraging the use of coal as a principal fuel (in lieu of gas or oil) by the electric utility and industrial sectors. Combustion residues from coal-fired power plants—fly ash, bottom ash, boiler slag, and fuel-gas-desulfurization (FGD) sludge—are currently exempted from the RCRA (Resource Conservation and Recovery Act), which requires the EPA to promulgate regulations for the disposal of hazardous and nonhazardous wastes.[38]

On the basis of chemical origin, the EPA categorizes the wet waste streams for the steam-electric power generating point single source category as follows:

1. Once-through cooling water
2. Recirculating cooling system blowdown
3. Fly ash transport discharge
4. Bottom ash transport discharge
5. Metal cleaning wastes
   Air preheater
   Fireside wash, etc.
6. Low-volume wastes
   Boiler, evaporator, softener blowdowns
   Drains, sanitary wastes, etc.
7. Ash pile runoff

---

[37] *Monthly Power Plant Report*, Energy Information Administration, Form E1A-759 (1986), Washington, D.C.

[38] "Impacts of Proposed RCRA Regulations and Other Related Federal Environmental Regulations on Fossil Fuel-Fired Facilities," *Engineering Science*, DOE/ET/13543-2316 (Washington, D.C.: U.S. Department of Energy, 1987).

---

8. Coal pile runoff
9. Wet flue-gas cleaning blowdown

Mean and maximum values (discharge flow rates per installed capacity) of these waste streams for coal-fired power plants are tabulated in Table 5.5. It should be noted that cooling water and wet ash handling systems and flue-gas scrubber process produce the major discharge volume.

The mean discharge flow rate per installed capacity from once-through cooling water systems is approximately 900 times as much as that from recirculating cooling water systems. However, the effluent from once-through

Table 5.5   Wastewater Discharge Flow Rates for Coal-Fired Power Plants

Waste Streams	Mean Value (GPD/MW)[a]	Maximum Value (GPD/MW)
Cooling System		
Once-through	1,140,619	55,430,000
Recirculating	2,937	63,056
Wet Ash Handling		
fly ash pond overflow	3,808	16,387
bottom ash pond overflow	3,881	38,333
Flue-Gas Scrubber		
Solids pond overflow	3,973	195,000
Blowdown	811	8,824
Boiler Blowdown	148	3,717
Evaporator Blowdown	126	8,292
Other Streams[b]	288	6,600

[a]Million gallons per day per megawatt of installed capacity.
[b]Other streams represent the total waste discharge from ion exchange softener, spent regenerant blowdown, filter backwash, clarifier blowdown, lime softener blowdown, air preheater and boiler fireside wash waters.

Source: D. M. Costle, *Development Document for Proposed Effluent Limitations Guidelines, New Source Performance Standards, and Pretreatment Standards for the Steam Electric Point Source Category*, EPA 440/1-80/029-b (Washington, D.C.: September 1989).

systems is relatively clean, and does not need treatment prior to discharge. In Table 5.5, the waste streams associated with the "Other Streams" processes may not exist in every coal-fired power plant. A general description of systems handling major wastewater streams follows.

Cooling Water Systems. In a steam-electric power plant, cooling water absorbs the heat that is liberated from the steam when it is condensed to water in the condensers. Depending on its size, and location and the availability of a water body, a power plant may have either of the cooling water systems, once-through or recirculating.

1. *Once-through cooling water system.* In a once-through cooling water system, the cooling water is withdrawn from the water source, passed through the system, and returned directly to the water source. Discharge flow rates from such a system in coal-fired power plants may reach up to 55.4 MGD/MW. In the United States, about 65% of all power plants have once-through cooling water systems.
2. *Recirculating cooling water system.* In a recirculating cooling water system, the water is withdrawn from the water source and passed through the condensers several times before being discharged to the receiving water. After each pass, the heat is removed from the water by any of three major methods: cooling ponds or cooling canals, mechanical draft cooling towers, or natural draft evaporative cooling towers. Discharge from such a system in coal-fired power plants may reach up to 63,057 GPD/MW.

Ash-Handling Systems. The chemical compositions of both types of bottom ash, dry and slag are quite similar. The major species present in bottom ash are silica (20% to

60% weight as SiO_2), alumina (10% to 35% weight as Al_2O_3), ferric oxide (5% to 35% weight as Fe_2O_3), calcium oxide (1% to 20% weight as CaO), and other minor amounts of metal oxides. Fly ash generally consists of very fine particles. The major species present in fly ash are silica (30% to 50% weight as SiO_2), alumina (20% to 30% weight as Al_2O_3), and other species including sulfur trioxide, carbon, boron, and so on. Distribution between bottom ash and fly ash varies depending on the type of boiler bottom.

Typically, the bottom ash to fly ash ratio is 35:65 for wet bottom boilers, and 15:85 for dry bottom boilers.[39]

Wet ash handling (sluicing) systems produce wastewaters that are currently either discharged as blowdown from recycle systems or discharged directly to receiving streams in a once-through manner. In a coal-fired power plant, wet ash handling system discharges may change, reaching up to 16,387 GPD/MW for fly ash ponds, and 38,333 GPD/MW for bottom ash ponds.

Flue-Gas Desulfurization Processes. In the lime or limestone flue-gas desulfurization processes, SO_2 is removed from the flue-gas by wet srubbing from slurry of calcium oxide (lime) or calcium carbonate (limestone). The principal reactions for absorptions of SO_2 by slurry are as follows:

Lime: $SO_2 + CaO + \frac{1}{2}H_2O \rightarrow CaSO_3 \bullet \frac{1}{2}H_2O$

Limestone: $SO_2 + CaCO_3 + \frac{1}{2}H_2O \rightarrow CaSO_3 \bullet \frac{1}{2}H_2O + CO_2$

Oxygen absorbed from the flue-gas or surrounding atmosphere causes the oxidation of absorbed SO_2. The

---

[39]D. M. Costle, *Development Document for Proposed Effluent Limitations Guidelines, New Source Performance Standards, and Pretreatment Standards for the Steam Electric Point Source Category*, EPA 440/1-80/029-b September 1980 Washington, D.C.

calcium sulfite formed in the principal reaction and the calcium sulfate formed through oxidation are precipitated as crystals in a holding tank. There is the potential to use calcium sulfite in manufacturing cement within the proposed complex.

CEMENT MANUFACTURING PLANTS

Portland cement is made by mixing and calcining calcareous and argillaceous materials in the proper ratio. Table 5.5 summarizes the relative amounts of raw materials consumed in the production of Portland cement. It can be observed that limestone represents the major amount of raw material consumed. The composition of regular Portland cement includes approximately 2.9% compound calcium sulfate ($CaSO_4$), as calculated from oxide analysis.

Unit processes involved in cement manufacture include storage and mixing of raw materials, drying, grinding and crushing, calcining, clinker storage, finishing additives and ball milling, and packaging for delivery. A schematic diagram of a typical rotary steam kiln boiler and an isometric flowchart for the manufacture of Portland cement are shown in Figures 5.18 and 5.14, respectively. For each 100 barrels of finished cement by the dry process, 297,872 Btu's of fuel, as well as 6.4 kilowatt-hours of electricity, and 8 gallons of water, are required. Also required are 132 pounds of limestone, 33 pounds of shale, and 4.3 pounds of gypsum.[40]

CONCRETE BLOCK MANUFACTURING PLANTS

The aggregates used for mortars and concretes can be conveniently divided into dense and light weight types. The first class includes all the aggregates normally used in mass and reinforced concrete, such as sand, gravel, crushed rock, and slag. The lightweight class includes

---

[40]Shreve and Brink, p. 162.

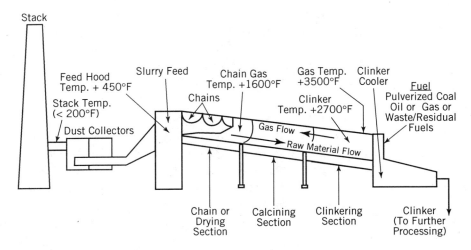

Figure 5.18. Schematic Diagram of a Typical Rotary Steam Kiln Boiler

pumice, furnace clinker (or "cinders" in the United States), foamed slag, expanded clay, shale, and slate. Among the lightweight aggregates, cinders are least expensive but variable in quality.[41] Table 5.6 summarizes some characteristics of concrete made with lightweight aggregates.

Environmentally Balanced Industrial Complexes

An Environmentally Balanced Industrial Complex has been proposed for a coal-fired power plant. A schematic flow diagram for the proposed complex is shown in Figure 5.19. The complex consists of three types of industrial plants, including a coal-fired power plant and two ancillary power plants: cement and concrete block manufacturing plants. The coal-fired power plant has the following waste streams:

---

[41]F. Meashan, *The Chemistry of Cement and Concrete*, 3d ed. (Surrey, England: Edward Arnold Publishers, 1970).

174

Table 5.6 Characteristics of Concrete Made from Light Aggregates

Aggregate	Density of Dry Concrete (lb/ft³)	Compressive Strength (lb/in²)	Drying Shrinkage (%)	Thermal Conductivity (Btu-in/ft²h°F)
Pumice	50	600	0.04–0.08	1–2
Clinker	60–95	300–1000	0.04–0.08	2.5–4
Expanded clay or shale	60–75	800–1200	0.04–0.07	2.3
Sintered pulverized fuel ash	70–80	600–1500	0.04–0.07	2–3
Foamed slag	60–95	300–1000	0.03–0.07	1.5–3

Source: F. Meashan, *The Chemistry of Cement and Concrete*, 3d ed. (Surrey, England: Edward Arnold Publishers, 1970).

1. Recirculating system cooling water blowdown
2. Boiler evaporator blowdown
3. Fly ash discharge
4. Bottom ash discharge
5. Flue-gas discharge

Cooling water blowdown, boiler blowdown, and evaporator blowdown are determined to be the major waste streams from a coal-fired power plant. These streams will be directed to the kiln steam boiler in the cement manufacturing plant, as shown in Figure 5.19.

Sulfur dioxide, which is released during the combustion of coal, will be scrubbed with a lime/limestone slurry. The calcium sulfate formed after oxidation will then be utilized in the cement manufacturing plant as a cement additive.

Waste dust from the kiln steam boiler in the cement manufacturing plant, as well as fly ash and bottom ash formed during the combustion of coal in the power plant, will be transported to the concrete block manufacturing

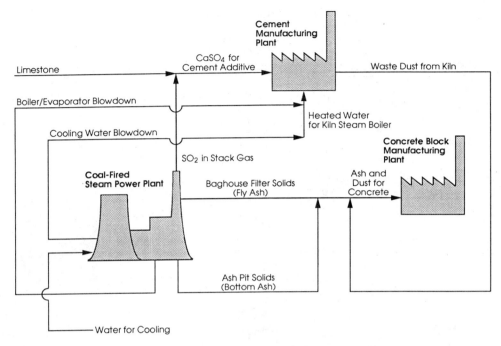

Figure 5.19. Schematic Flow Diagram of an EBIC for the Power Plant Industry

plant. These solid wastes will be utilized in the production of concrete blocks.

STEEL MILL–FERTILIZER–CEMENT COMPLEXES

A complex centering on an integrated steel mill is a necessity because of the vast amounts and varied types of contaminants emanating from a plant of this type. Moreover, a steel mill is ideally suited for a balanced complex since it typically operates within a multiplant center.

Integrated steel mills are actually five separate industrial plants in one, consisting of (1) a coke plant, (2) an iron ore reduction plant, (3) a steel production facility, (4) a hot rolling mill, and (5) a cold rolling mill. The predominant wastes originate from the coke and steel production plants. Dust, slag, and iron also come from the other plants.

Troublesome waste products include ammonia, cyanide, phenol, heat, and acidic ferrous sulfate or chloride pickle liquor. Steel mills also use huge volumes of water, mostly for cooling and quenching, which produces like volumes of air, water, and solid contaminants. Steel manufacturing has developed a worldwide reputation as one of the most polluting industries of modern times. The mills require so much land area and employ so many people that their location in a separate industrial complex would be a natural development. Fertilizer and building material plants are likely candidates for auxiliary industries. Such a complex is proposed, as presented in Figure 5.20.

In this complex we plan to produce five products for external sale, manufactured by the following plants:

1. Fertilizer
2. Cement
3. Steel
4. Hydrofluoric acid and/or etched glass
5. Building block

Major raw materials required for the complex include phosphate rock, sand, sulfuric acid, hydrochloric acid, and soda ash, in addition to water and power. The fertilizer plant uses the phosphate rock; the cement plant, building block, and glass plants use the sand; the fertilizer and steel mills also use the sulfuric acid; and the glass plants uses soda ash.

Gypsum waste from the fertilizer plant is reused by both the cement and the building block plants. *Dust waste* from the latter two plants is reused. *Slag solid waste* from the

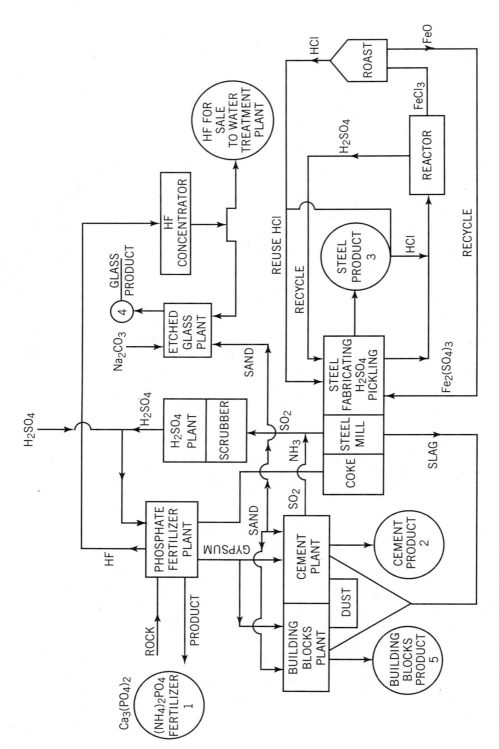

Figure 5.20. Steel Mill–Fertilizer–Cement Complex

steel mill is also used by the cement and block plants. *Sulfur dioxide gaseous waste* from the cement plant is converted to sulfuric acid and reused in both the steel mill and the fertilizer plant. *Iron solid waste* from the rolling mills (steel fabricating) is recycled in the furnaces to produce more steel. *Hydrofluoric acid* is recovered from the gaseous effluent from the fertilizer plant, concentrated, and either reused in the etched glass plant or sold directly to external water plants for fluoridating water supplies. Steel mill *pickling liquor waste* (largely $Fe_2(SO_4)_3$) is reacted with hydrochloric acid to produce both sulfuric acid, which is recycled in the pickling operation, and ferric chloride, which is roasted to make reusable hydrochloric acid an iron oxide. The latter is returned to the furnaces to make additional steel. No gaseous, liquid, or solid waste leaves the complex to contaminate the environment.

Phosphate fertilizers and cements are manufactured as described in the earlier discussion of the fertilizer-cement complex in this chapter. Etched glass is manufactured by melting a mixture of soda and sand with minor additives such as salt cake, powder, coal, and lime. The tank or pot furnace is maintained at the very high temperature of 2,600°F. Melted glass is withdrawn from the furnace, usually in spools, and shaped and cut to specifications as it cools. Cullet, or broken or cut pieces of glass is wasted during the forming process. The cullet is returned to the furnaces for remelting into new glass. In recent years (mainly beginning in the 1980s) steel production has evolved into (1) integrated producers, (2) minimills, and (3) specialty steel mills. Integrated steel mills, as described earlier, start with iron ore and coal and end up with steel of many shapes, which amount to about 70% of the United States market. This type of steel production results in the most waste, since it uses a coke oven and blast furnace prior to the basic oxygen furnace (BOF) to evolve molten steel. The minimills reprocess scrap steel, usually into some low-quality products. In these mills, coal and iron ore along with scrap are fed directly into the oxygen furnace and then into an electric furnace to produce molten steel. Thus, the major coke oven wastes and blast furnace dust and slag are eliminated. Specialty mills are similar to minimills but smaller, and manufacture more costly steels. In these mills mixtures of reduced iron and scrap steel are added, along with lime and some flux to the electric furnace to produce molten steel. No coal or raw iron is needed. Both the minimills and the specialty mills are using relatively new technology for refining and casting, two of which are direct iron making and continuous casting.

Direct iron making is a one-step production of molten iron from iron ore and coal. Major reductions in capital and operating costs, as well as waste treatment costs, result, as compared to those incurred by integrated production mills. Further, molten steel may be produced in one step (as used in integrated mills) and transformed to finished products. When this is possible, capital, labor, and energy costs are reduced, along with wastes from other processes. The molten iron produced by direct iron making is similar to that made in the blast furnace of the integrated mill. Here both the blast furnace and the coke oven are eliminated in favor of direct production in the modified basic oxygen furnace. Coal serves the dual purpose of a reducing agent and a fuel to melt the iron. However, natural gas (where economically available) can also serve to provide the CO/H_2, reducing the amount of air needed in the process.

Continuous casting, another recently developed steel production improvement, manufactures steel sheets a few millimeters thick, rather than rolling them laboriously from ingots or slabs.

PLASTIC INDUSTRY COMPLEXES

Manufacturing Process

Plastics and resins are chainlike structures known chemically as polymers. Polymers are synthesized mainly by adding a free radical initiator and a modifier to the monomer (the building block of the polymer). Although not a great deal of wastewater arises from the polymerization process, more results from synthesizing the original monomer.

As long ago as 1967, the U.S. Department of the Interior classified plastics in nine categories, but we are primarily interested in only two predominant types: polysterene and polyolefins. In 1967, these types made up 46% of the total plastics and resins produced.[42]

Polysterene's combination of physical properties and ease of injection molding and extruding makes its use desirable. The crystal-clear product has excellent thermal and dimensional stability, high flexural and tensile strength, and good electrical properties.[43] Styrene monomers (or mixtures) are purified by distillation or caustic washing to remove inhibitors. The purified raw materials, together with an initiator, are fed to a stainless steel or aluminum polymerization vessel, jacketed for heating and cooling and containing an agitator. Polymerization of the monomer is carried out at about 90°C to approximately 30% conversion to create a syrupy mass. Water is used during this stage as a heat exchange medium and is recirculated without contamination. Then the syrupy mass is transferred to suspension-polymerization reactors containing water and proprietary suspending and dispersing agents. These reactors are usually jacketed, and the contents stirred in stainless steel vessels.

---

[42]Nemerow and Dasgupta, *Industrial and Hazardous Wastes*, 568.
[43]Ibid., pp. 567–575.

The syrupy mass is broken into droplets by means of the stirrer and held in suspension in the aqueous phase. Temperature is a critical variable in further polymerization of the product. The polymer is then sent to a blowdown tank where any unreacted monomer is stripped. The stripped batch is then centrifuged, and the polymer product is filtered, washed, and dewatered. A flowchart of this manufacturing process is presented in Figure 5.21.

Reaction water (suspension medium) and wash water are the two significant sources of wastewater from the manufacturing of polysterene. They are shown in Figure 5.21 as wastes A and B. About 1.5 gallons of water, excluding cooling water, is used and wasted for each pound of polysterene produced. These wastes are not very polluting; they contain small amounts of catalyst and suspending agents used in suspension polymerization, and heated water (120°F to 180°F).

The catalysts are generally of the peroxide type, and the suspending agents may be methyl or ethyl cellulose, polyacrylic acids, polyvinyl alcohol, or miscellaneous compounds such as gelatin, starch, gum, casein, zein, and alginate. Inorganic materials such as $CaCO_3$, $Ca_3(PO_4)_2$ talc, clay, and silicate may also be present in effluent reaction water. No plants employing typical technology have waste treatment units; instead, 90% discharge these wastes into municipal sewers.

Polyolefins (polyethylenes) are composed of many different molecular weights from waxes of a few thousand to molecular weights of several million. Polyethylene is used for film and sheet, injection molding, blow-molded bottles, cable insulation, coatings, and other products.

There are two processes for manufacturing polyethylene: one for a low-density and the other for high-density product. Both start with ethylene as the raw stock material. Heat, pressure, catalysts, and solvents are reacted

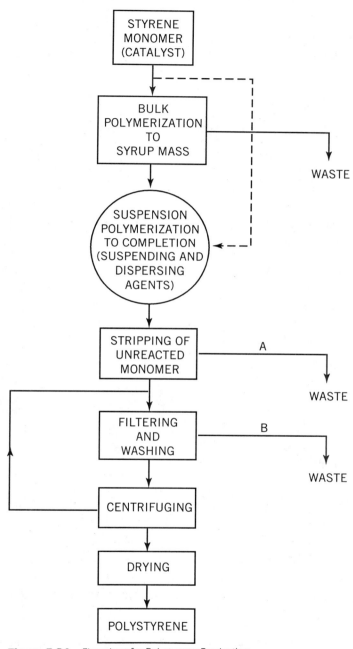

Figure 5.21. Flowchart for Polystyrene Production

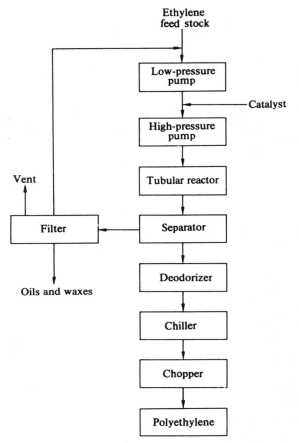

Figure 5.22. Tubular Reactor Process for Low-Density Polyethylene Production

with the ethylene, and then the product is purified by various unit operations as shown in Figures 5.22 and 5.23. Both processes produce little wastewater. However, potential hazards that may generate water-borne waste are improper operation, spills, and washdown of equipment and facilities. Typical process wastewaters contain a BOD of less than 10 ppm.[44]

[44]"The Cost of Clean Water", Vol. III Industrial Waste Profile No. 10: Plastic Materials and Resins (Washington, D.C.: U.S. Dept. of the Interior, 1967).

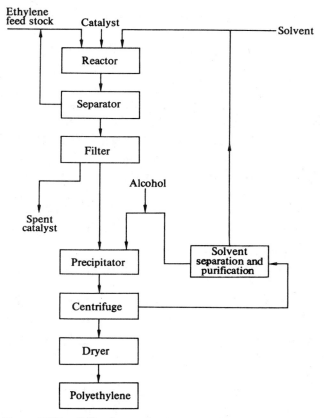

Figure 5.23. Phillips Process for High-Density Polyethylene Production

The Ultimate Plastic Problem

According to S. P. Sherman,[45] the United States produces 60 billion pounds of plastics annually, and sales of plastic products exceed $150 billion a year. By the year 2000, the U.S. output may reach 76 billion pounds. Thayer[46] expects plastic waste to reach 38 billion pounds per year at that

[45]S. P. Sherman, "Trashing of a $150 Billion Business," *Fortune* (August 1989): 96.
[46]Thayer. Chemical and Engineering News Bureau (Northeast), 1989.

time and to account for 10% of the total municipal solid waste.

Since plastic production wastes are minimal as far as contaminants are concerned, little or no waste treatment is done at present at plastics manufacturing plants. This industry is rather unique in that the major wastes and environmental costs occur during and following the use of plastic products, which makes plastic waste treatment very difficult.

Until the late 1980s most plastic products were disposed of in municipal solid wastes and were given the same treatment as these wastes. Landfilling and incineration were the predominant systems for plastic waste treatment in solid wastes.

N. P. Cheremismoff and P. N. Cheremismoff report that "the option of landfill as a disposal method is rapidly diminishing."[47] For example, the number of legal landfills declined from a 1976 figure of 18,000 to 9,000 in 1989. Not only are landfill sites diminishing, but costs and limits for disposal of plastic materials are increasing. Plastics degrade slowly in soils relative to other ingredients in municipal solid wastes. Since we are rapidly approaching a crisis in the treatment/disposal of plastics, we must seek alternate and new methods.

The best choice, according to S. P. Sherman,[48] is waste-to-energy incineration. Most plastics ignite as easily as natural gas and emit CO_2, NOx's, and H_2O vapor. However, some plastics, mainly polyvinyl chlorides (PVCs), can adversely affect the air environment. These PVCs, unless burned at greater than 1,200°F to 1,600°F, will emit dioxins. Even at these temperatures they will evolve hydrochloric acid, which is corrosive to metals and antag-

---

[47]N. P. Cheremismoff and P. N. Cheremismoff," The Plastic Waste Problem Report," *Pollution Engineering* (August 1989): 58–68.
[48]Sherman, "Trashing of a $150 Billion Business," p. 96.

onistic to humans who come in contact with the acid. Some other plastics contain cadmium or other heavy metals, which remain unburned in the incinerator ash. Such metals are toxic when they enter the water environment through landfill leachates. One advantage of incinerating plastics is the high heat energy released from the burning: about 16,000 BTUs per pound. Some waste-to-energy type plants, which are rather expensive to build (about $100,000/ton/day), are described in Chapter 16 of *Industrial Solid Wastes*.[49]

Other potential problems with incineration as a method of treatment/disposal for plastics include enhancement of the "greenhouse effect" and ash disposal. The burning emits CO_2, which keeps the heat energy rays (infrared) from escaping the earth. The subsequent heating may alter our earth's climatological condition, affecting agricultural output and ocean levels. However, there is still much controversy about any true CO_2 increase in our atmosphere. Even so, the small increase in CO_2 may also stimulate increased tree and plant growth. These green growths will evolve oxygen into our atmosphere and oceans to aid in their purification.

The ash resulting from incineration may constitute about 10% of the waste burned, contain heavy metals, and be expensive to dispose of by normal processes. Some progress has been made in using ash as an additive in concrete products meeting utilization specifications.

Another possibility for eliminating the plastic disposal problem is using waste plastics to make other useful products. Rubbermaid Company, for example, makes trash containers in Winchester, Virginia out of reused plastic chips.[50] Lehrman owns BTW, a company that reprocesses used plastics to make woodlike fence posts,

[49]N. L. Nemerow, *Industrial Solid Waste* (New York: Ballinger, 1984).
[50]Sherman, "Trashing of a $150 Billion Business, p. 96.

car stops, and picnic tables.[51] Main reprocessing units include pulverizers, extruders, and other pressure form machinery. The cost of reused plastic varies from $0.23 to $0.40 per pound, almost as much as virgin resins.[52] Prices, however, do not present the problem that collection and separation of the different plastics cause. Recycled polyethylene can also be used for carpet fibers, and, polystyrene is used for a variety of durable goods such as office supplies, hair accessories, cafeteria trays, license plate holders, and loose-fill packing material.

As of 1991 it appears that the major deterrent to recycling used plastics for reuse in other products is the difficulty of finding a convenient market near the source of the waste. This market must be able to reuse a sufficient quantity of plastic at an equitable raw material price. All of these factors limit recycling into other products as a universal method within the industry.

At this point in the solution to the problem, we suggest, once again, the use of a plastic industry environmentally balanced complex. However, in this case we propose merging only two facets of the plastic industry: virgin plastic manufacturing and recycling plastic collectors. Such mergers would result in recycled plastic materials in the immediate area being reprocessed by the original plastic manufacturer into plastic products similar to the originals. An example of such a complex is shown in Figure 5.24.

Some limits on quantity and quality of recycled plastics and additions of processing equipment undoubtedly will be necessary to optimize the complex. However, the merger appears worthwhile to both industries. The plastic

[51]Joe Kollin, "Recycling Firm Puts Old Plastics to Use," *Fort Lauderdale Sun Sentinel* (10 July 1991) News Section, 4.
[52]"Profitability Problems Plague Plastics Recycling" (The Chemical Environment), *Chemical Business* (March 1990): 34.

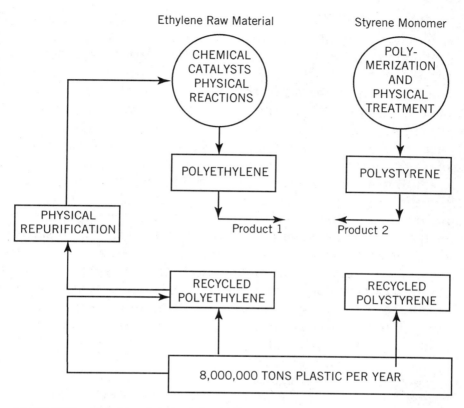

Figure 5.24. Plastic Manufacturing Industrial Complex

manufacturer will have lower raw material costs and recyclers will not have typical marketing problems with the final product. In addition, as usual, no wastes result from the complex and a considerable plastic solid waste problem will have been removed from the environment.

CEMENT, LIME, AND POWER PLANT COMPLEXES

In many cases it is possible to combine more than two industries in a complex to more fully utilize all waste products and optimize production costs. Such a situation exists in the manufacturing of cement, lime, and power.

We have already discussed the separate production of cement and coal-fired power production (see Figure 5.19). The reader is referred to these earlier sections for review of their manufacturing materials processes and wastes. In this section we will present the background of lime manufacturing and illustrate how it would work into the complex concept.

Lime production is an ancient industry. Today lime and limestone are employed in more industries than any other natural substances.[53] Lime itself is used for medicinal purposes, plant and animal feed, insecticides, gas absorption, precipitation, dehydration, and caustisizing. It is used as a reactant in paper making, dehairing hides, manufacturing high-grade steel and cement, water softening, recovery of by-product ammonia, and the manufacture of soap, rubber, varnish, refractories, and sand-lime brick. Lime is also an essential ingredient in mortar, plaster, and soil additives.

There are several types of limes, varying with the lime content, water content, and specific use requirements. All types originate with limestone, hence lime manufacturing is usually done nearby limestone mining deposits.

The carbonates of calcium and magnesium are obtained from deposits of limestone, marble, chalk, dolomite, and oyster shells. Quarries are selected that yield a rock consisting of low amounts of silica, clay, or iron and a naturally high concentration of calcium, magnesium, or both. Impurities can adversely affect the desired hydraulic qualities of the resulting lime. Both overburned and underburned limestone leave undesirable lumps in the product lime.

---

[53]Shreve and Brink, 166.

Considerable energy (power) is required in this industry for blasting limestone out of the mines, for transporting and sizing of rock, and for burning (calcining) it. Calcining requires 4.25 million Btu's/ton of lime produced, and subsequent hydration liberates 15.9 kilocalories. The volume of rock declines during calcining and swells during hydration. The amount of coal required for calcining varies from 1 pound to 3 1/3 pounds (depending on the kiln type) per 3 1/4 pounds of lime produced. Calcining takes place at 1,200°C to 1,300°C.

The sequence of actual operations in manufacturing lime is (1) *blasting* (the limestone in the quarry), (2) *transporting* the stone to the plant, (3) *crushing* and *sizing* of stones, (4) *screening* to remove small (< 4 in) and large (> 8in) stones, (5) *moving* the uniform-size stones to a vertical kiln (large size) or rotary kiln (small size), (6) *taking fines* to a pulverizer to produce powdered limestone for agriculture uses, (7) *burning* the limestones in vertical kilns (to give lump lime) or in horizontal rotary kilns (to make fine lime), (8) *packaging* of finished lime in barrels or drums or sending it to a hydrator to make hydrated lime, and (9) *packaging* of slaked lime in bags.

A typical schematic flowsheet is shown in Figure 5.25. An existing complex of this type is already in operation in Brooksville, Florida, as shown schematically in Figure 5.26. The complex is claimed to be "the world's first combination of pulverized coal/fluidized bed combustion boiler producing lime and electric power, integrated with a cement plant. . . . Nowhere else in the State [Florida] is there a facility that's using the waste material from the mining, in the case of limestone fines which could not be sold otherwise, and making major products like Portland cement and lime."[54]

[54]A. R. Lawhome, "Integrated Cement, Power and Lime Facility Obtains Maximum Cost Efficiency," *The Florida Specifier* (December 1989), 1.

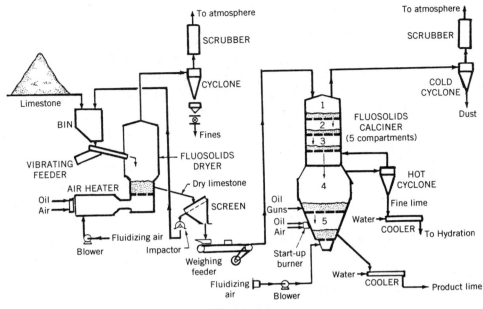

Figure 5.25. FluoSolids System (Dorr-Oliver, Inc.)

The Brooksville plant will produce about 600,000 tons of cement a year and 330,000 tons per year of chemical lime. The lime or CaO will be used for gas desulfurization, acid pond naturalization, agriculture or in building products. The limestone quarry plant "generates about a ton of waste product for each ton of usable stone. . . . The limestone fines are not suitable as an aggregate or as a stable fill material, but do contain large quantities of calcium and silica.

"This composition is required in raw material in the manufacture of cement. . . . The plant's energy efficient co-generative design allows the simultaneous manufacture of lime and cement, which, in turn, reduces operational costs of the electrical power plant. Hot air, which comes from the cooling plant cement clinker, is used as combustion air for the power plant boiler. Waste heat in the form of cement produced by the generator is used to dry the limestone at the cement and lime operations. In addition to lower power costs, fly ash from the combustion of coal in the power plant provides additional iron and aluminum—two cement raw materials that are not present in adequate quantities in the limestone fines.

A state-of-the-art computerized control room, manned 24-hours a day, supervises every aspect of the plant's production facilities. Highly trained personnel monitor production quality to ensure that finished products meet or exceed strict specifications. A chemical analysis department repeatedly tests lime and cement samples for mineral content and consistency.

Initial engineering and design for the cement power and lime cogeneration project got under way in 1982, and production began in 1984. Various components of the facility, including the cement plant, have been in operation since about 1988.

Design for the 125 megawatt coal-fired plant was provided by LDP (Larramore, Douglass and Popham, Inc.) of Chicago. The power produced at this facility will be sold under a long-term contract to Florida Power & Light Company."[55]

WOOD (LUMBER) MILL COMPLEXES

Sawmill and planing mills (Standard Industrial Classification code 2421) produced 162 tons of solid waste per employee per year. Small producers of lumber and wood products also generated 16,083 cubic yards of solid waste per firm per week, or 836.33 cubic yards per firm per year.[56] This was contributed by 17.247 employees per firm, for an annual discharge of 48.492 cubic yards per employee.

---

[55]Ibid.
[56]N. L. Nemerow, *Industrial Solid Wastes*, 176, 190.

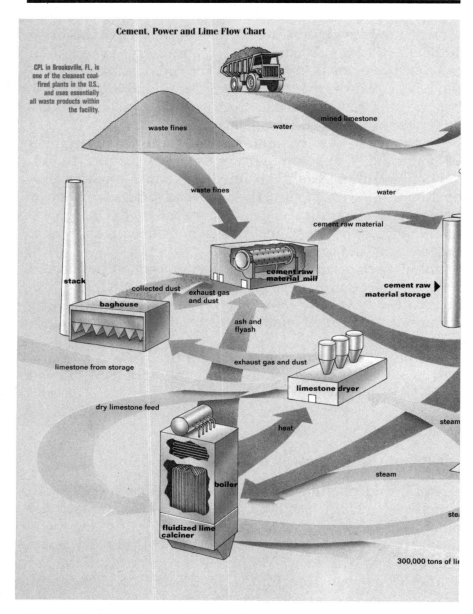

Figure 5.26. Cement, Power, and Lime Flowchart

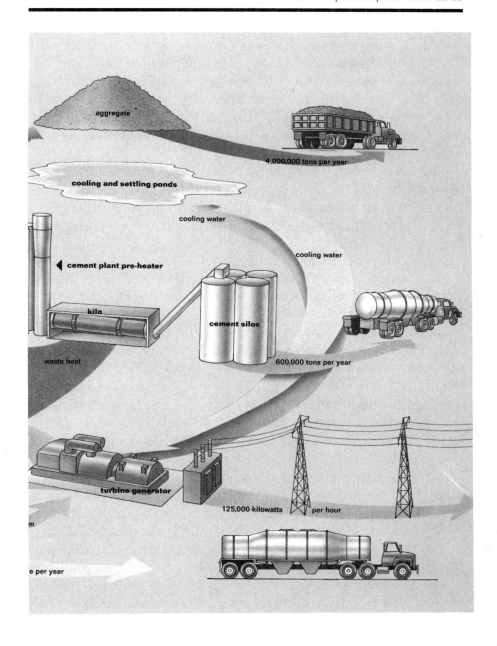

aggregate

4,000,000 tons per year

cooling and settling ponds

cooling water

cooling water

cement plant pre-heater

kiln

cement silos

waste heat

600,000 tons per year

turbine generator

125,000 kilowatts   per hour

m

e per year

Rough lumber from trees is brought to the lumber mill and sawed and planed to appropriate lengths and widths, largely for sale in the housing market. Planing removes the tree bark, and saws produce sawdust in addition to finished lumber product. Most sawmill waste (sawdust and bark) has typically been burned or reclaimed as soil conditioner.[57]

The burning of these solid wastes, while not requiring an outside source of energy, contributes potential air pollutants of unburned carbon and ash. Soil conditioning with these same wastes requires preparation, transporting, and locating a suitable market, all of which cost lumber mill owners time and money.

A potential solution to the problem of sawdust and bark wastes originating from lumber mills is a EBIC in which these wastes are used directly to produce other products. Such a complex is shown schematically in Figure 5.27.

In this complex the plant will pyrolyze the wastes (sawdust and wood chips) into oil vapors. The vapors are condensed to make fuel oil or other chemicals (product 3). The remaining gases are burned to generate electricity (product 2). Finished lumber, naturally, is the prime product of the complex (product 1). Nonburnable gases such as CO_2, CO, and NO's are also obtained from the condenser/pyrolyzer and used to stimulate the growth of algae in a separate unit operation. This takes place in a greenhouse in order to utilize natural sunlight. The algae serves as food for small seafood such as shrimp which, along with some algae/herbs, are sold as food products (product 4).

Such a complex has been suggested[58] to convert 70% to 90% of the wood waste into salable products; at the same

---

[57] Ibid., 180.

[58] "Process Converts Wood Waste into Chemicals" *Pollution Engineering* (March 1991): 45.

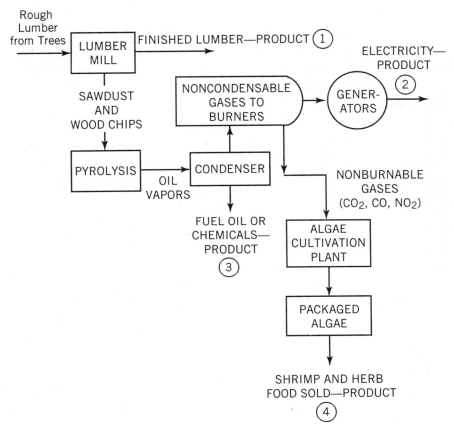

Figure 5.27. Lumber Mill Complex—Four Products

time virtually no pollutants are emitted. There are also possibilities for converting bark and sawdust into other useful products, such as resins and tanning agents from bark and product fillers from sawdust. These would need exploration by individual cases.

POWER PLANT–AGRICULTURE COMPLEXES

Electric power and food are two of the vital necessities for human life. Without electricity our society would revert back to ancient times and, without food would starve to death. It is imperative that we preserve both industries and produce their products as efficiently as possible. By combining their operations in one complex, we can accomplish two objectives: efficient production and a pollution-free environment.

A possible configuration of consolidating electric power and food production is shown in Figure 5.28. In this complex, a low-sulfur fossil fuel power plant produces three main wastes: (1) heated water, (2) flue-gas fly ash, and 3) boiler residue ash. These wastes are described in detail in the earlier section "Fossil-Fueled Power Plant Complexes." The heated cooling water is reused within the complex to enhance the growing of fish. Fish will metabolize food faster and thereby grow at a greater rate if the temperature is raised—especially in colder climates.

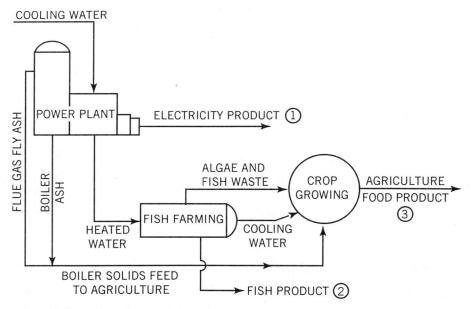

Figure 5.28. Power Plant–Agriculture Complex

Both ashes from the flue gas and boiler residue are fed to the agriculture growing area to enhance the soil character. The minerals in these ashes will increase crop growth and yield more food per acre. At the same time, no ash wastes will reach the outside environment.

As shown in Figure 5.28, three products will be sold outside the complex: (1) electric power (2) fish and (3) agricultural food crops. Algae and fish wastes, produced in the fish-farming waters, will also be reused as fertilizer for the agricultural crop. No pollution will reach the air, water, or land environments outside this complex—especially if the power plant uses low- or no-sulfur fuel. This would be possible by using only low-sulfur coal or oil or natural gas as fuel.

POWER PLANT–CEMENT–CONCRETE BLOCK COMPLEXES

In another three-plant complex we have the potential of eliminating all air and water pollutants from three essential construction-related industries. To make cement and concrete block it is necessary to provide, among other things, power. Power is also necessary for the homes and the buildings using the cement and block products.

If all three plants are located in the same industrial complex, the marketing of their essential cement, power, and blocks become accessible to the area nearby to the complex. The manufacture of cement is described in the earlier section of this chapter "Fertilizer–Cement Complexes," and fossil-fuel power production is detailed in the next section, "Fossil-Fuel Powered Complexes."

To make concrete blocks requires stone, sand, cement, and water, and any additives that give the block product strength, rigidity, durability, lightness, and bulk, all of which must not result in excessive cost per block. Blocks are usually 4 or 8 inches thick, 12 inches long, and 8 inches deep. They are manufactured by mixing or blending the raw materials with water, pouring the mixture into molds, applying pressure, and curing the preformed block in wet air or steam for a given period of time.

Figure 5.29 depicts an EBIC consisting of three industrial plants in the same location. In this complex, three main products are produced for sale and use within and outside the complex: (1) power, (2) cement, and (3) concrete blocks.

Solid wastes leaving the fossil-fuel power plant are boiler ash and stack fly ash. Both are captured and sent to the concrete block plant for use as additives in the blocks. Sulfur dioxide in the flue gas is scrubbed in limestone filters to form calcium sulfate ($CaSO_4$). The resulting gypsum material is sent to the cement plant to serve as a source of calcareous material in making cement. Plant

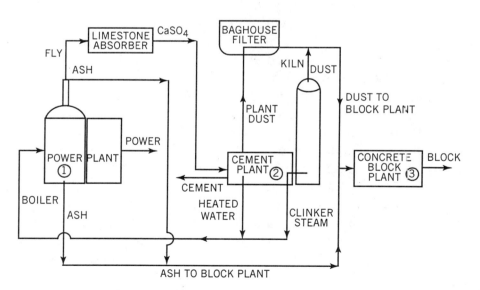

Figure 5.29. Power Plant–Cement Plant–Concrete Block Complex

dust and kiln dust from the cement plant are removed from the air by baghouse-type filters and also sent to the concrete block plant to be used as additives. Heated clinker cooling water from the cement plant is returned to the power plant to heat incoming boiler feed water.

No other major solid, gaseous, or liquid wastes leave these three plants. The surrounding environment remains uncontaminated by any effluent. At the same time, three products are being manufactured at a lower cost than would be required if the same plants were located some distances apart.

CANNERY–AGRICULTURE COMPLEXES

Canneries are in the business of producing food products for direct consumption by the purchaser. Under most existing conditions, they import fresh-grown fruits and vegetables from considerable distances. This not only results in added costs of transportation, but also increases spoilage and waste of raw materials during the travel and extra handling.

It would make good economic sense for canneries to grow their own raw materials within a complex. But since farming is an entirely different enterprise, coordination of farming with the cannery operation would be required. The land areas and their usage required in agriculture must be integrated with the relatively smaller areas of canning. Moreover, the two enterprises require workers with different backgrounds and abilities.

However, if the two plants (processing operations) could be combined in one complex, the entire system would be more cost-effective. In addition, the normal wastes produced by each plant would be utilized within the complex to eliminate any adverse environmental impact to the surrounding area.

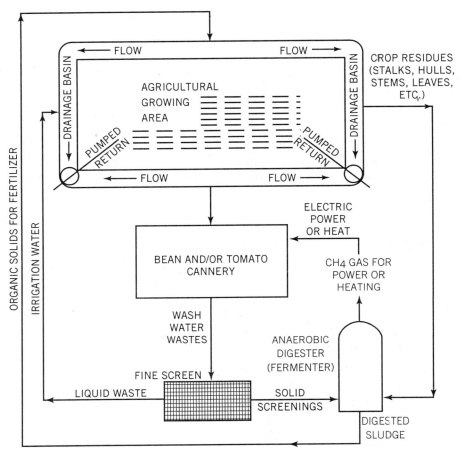

Figure 5.30. Cannery–Agriculture Complex

A typical configuration of such a complex is shown in Figure 5.30. In this complex, the farming industry would grow fruits such as tomatoes and peaches, and vegetables such as beans and carrots. These crops would be cut and or picked in the field and mechanically conveyed directly to the adjacent canning plant for processing. Waste cuttings, rot, and extraneous organic residues remaining after harvesting would be collected and trucked directly to the fermenter for digestion to methane gas.

Rainoff from the farming area, carrying excess or unused fertilizers and pesticides, would be collected in drainage ditches surrounding the growing area. This runoff, instead of polluting the area water environment, is collected and pumped back to the growing area through perforated aluminum pipelines. In this way, water is conserved, as well as potentially contaminating chemical phosphates and insecticides.

The wastewater from the cannery is screened and also returned to the farm growing area as a source of valuable water. The screenings are delivered to the farm to enhance the production of methane gas.

The fermenter digested sludge waste, which is distributed on the farmland crop-growing area increases the fruit and or vegetable yield because of its value as an organic fertilizer. The gas (CH_4) produced by the fermenter is used internally to heat the cannery buildings or sold externally to local power plants to produce power for the cannery or local homes of industrial workers.

No wastes leave the complex to cause adverse effects on the environment. Furthermore, canned fruits and vegetables are produced at a minimum cost.

NUCLEAR POWER–GLASS BLOCK COMPLEXES

One of the biggest environmental concerns facing the world today is the problem of waste disposal from nuclear power plants. Safety hazards appear to be waning as a result of better plant design, operation, and supervision. Low-level wastes have never caused serious environmental problems. Only high-level radioactivity resulting from replacing spent fuel rods has been of serious and ongoing to concern the Atomic Energy Commission (AEC) as well as to environmentalists. Nuclear energy, as a viable alternative to the use of coal or oil fossil fuel, is an attractive possibility for the future.

If the environmental consequences of nuclear energy production can be overcome, its acceptance and use will solve an enormous energy dilemma. This involves devising solutions for both the excessive heat releases in the cooling water and the dangerous radioactivity contained in the high-level fuel reprocessing wastes.

A possibility for abating both waste problems is the use of a balanced environmental complex, as shown in Figure 5.31. This complex contains plants producing three major products (1) electricity, (2) decorative glass blocks, and (3) a recreational boating facility.

Electricity here is produced by the nuclear power plant. The cooling water used to condense steam exhausting from the turbine contains excessive heat and high energy. It is used as a source of water for a white water recreational boating river. This river meanders toward an outlet water sink (lake, river, or ocean), losing its heat and potential energy as it flows downhill and around bends to the outlet. Boating obstacles are placed along the recreational path to enhance the challenge to its users. The outlet sink also serves as the source of cooling water intake for the condenser. Any potential low-level radioactivity leakage from the uranium fuel rods into the stream and, subsequently into the condensate, would be retained in the holding basin, whose effluent is continuously monitored before discharge to the recreational river.

The uranium fuel rods, when reprocessed, are removed to the adjacent cleaning plant where nitric acid and other cleaning agents are added to refurbish the rods. The high-level wastes are concentrated by evaporation before being sent to the block manufacturing plant. Here lime, silica, soda, and some coal are added, and the mixture is sintered

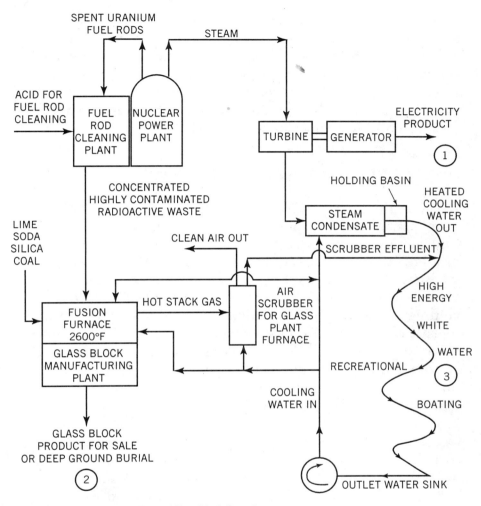

Figure 5.31. Nuclear Power–Glass Block Complex

at 2,600°F to form a molten glass product. The liquid glass is poured into blocks, cooled, and removed to be sold as a decorative building material or buried in deep, dry, underground locations for environmental security. The hot furnace gases are precooled by using some of the outlet sink water and then sent to a scrubber to remove any excess sulfur and carbon dioxide. The scrubber effluent is also returned to the white water recreational boating river. Thus, no radioactivity or heated water escapes from the complex. Furthermore, a recreational industry is supported, and electricity and glass block products are manufactured at lower total costs.

ANIMAL FEEDLOT–PLANT FOOD COMPLEXES

Large-scale livestock operations have removed animals from pasturage and now handle large numbers in small confinement areas (feedlots) where feed and water are brought to the livestock.[59] R. C. Loehr[60] reported that treatment and disposal of animal wastes-collected in the feedlots-are complicated by the nature of the wastes, the volume of wastes to be handled, the lack of interest by the livestock producer in waste treatment, and the proximity of a suburban population. These wastes are high in organic solids-both suspended and dissolved-BOD, and nutrients such as ammonia. Land disposal and anaerobic lagoons have been used to treat the wastes. As a result production costs have been increased to include waste treatments. It makes sense to eliminate these treatment costs by substituting ancillary industries to use the wastes.

Figure 5.32 shows a potential combination of plants to handle all feedlot wastes and, at the same time, produce other useful products. In this complex, the animal feedlot is the major plant and the hyacinth food plant and digester are ancillary plants. The feedlot produces mature

---

[59]Nemerow and Dasgupta *Industrial and Hazardous Wastes* 441.

[60]R. C. Loehr, "Effluent Quality from Anaerobic Lagoons Treating Feedlot Wastes *Journal of Water Pollution Control Federation* (February 1967) 39:384 (1967).

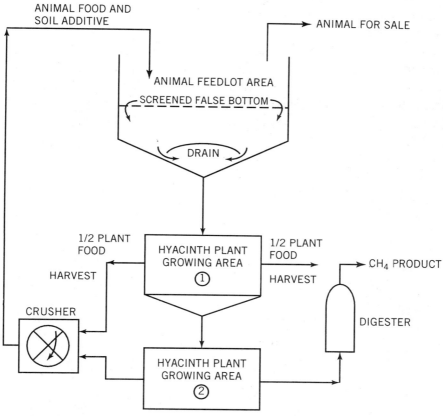

Figure 5.32. Feedlot–Food Production Complex

animals ready for sale in the open market, and hyacinth production makes plant food, partially for reuse as animal food and partially for feed for the fermenter. The fermenter makes methane (CH_4) gas for outside sale or for heating any of the complex buildings. A secondary hyacinth basin recovers settled sludge from the first basin and makes more hyacinth plants for similar usage.

All feedlot wastes are utilized in the complex to grow hyacinth plants in shallow growing basins in natural or artificial sunlight. Hyacinths are crushed and returned to the feedlot area as food for animals. The fermenter produces methane gas (CH_4) for digesting one-half of the hyacinth plants from the first growing area. The methane becomes the third product made in the complex.

By using a three-product industrial complex, we have provided animals with a feedlot area for a given period of time without creating any adverse outside environmental effects.

CHAPTER 6

SUMMARY

In this book I have attempted to show that there is a logical approach to attaining zero pollution from industry. Moreover, the logic leads to an even more obvious conclusion and an ultimate solution to realizing zero industrial pollution. That solution is simply to utilize all wastes produced by an industry as raw materials for another industy (or industries) located adjacent to it.

The need to obtain an industrial pollution-free environment is critical. In an expanding world economy, our fixed environmental resources will not tolerate any further deterioration of quality. We must produce more goods and, at the same time, create less pollution. Moreover, we must accomplish this at less cost. Society cannot stand a spiraling inflation fueled by increased industrial production costs. The "now accepted" practice of adding the cost of pollution control to the cost of a product must be discontinued. Instead, we must replace it with an ultimate "no added cost for waste treatment." To accomplish this end, and at the same time preserve the quality of our environment, is the crux of this book.

We can make a start at reaching zero industrial pollution by minimizing all our plant wastes. This is already being done in a variety of ways as industry closely surveys its production methods and resulting wastes.

Next we can—and have already begun—reusing within each industrial plant many of the wastes formerly sent to treatment facilities or (worse) to receiving waters, landfills, and air. Although reuse within our plants has its limitations as to product quality and quantity, the practice is alleviating the problem somewhat.

Where internal reuse is not possible or is limited by constraints, wastes can be sold and/or shipped to other industries for reuse either with and without waste modification. This solution involves some added costs and is not without its own limitations, such as requirements of marketing more products. However, this system also lessens the industrial pollution problem.

When it is difficult to identify reusers, industry can be assisted by using "market exchanges." These exchanges provide lists of buyers and sellers of wastes and will even arrange for sales to assist industries in some cases. Continuous arrangements betweeen buyers and sellers at dependable prices may be difficult to obtain, which can somewhat impede the use of this solution.

The ultimate solution to arriving at zero pollution, as described in the Chapter 5, is to locate compatible plants together in one complex so that one uses the others' wastes and no pollutants escape into the surrounding environment.

We have described the urgency and rationale for attaining zero pollution. The major concern in implementing the plan we propose is economics. It must be demonstrated that recycling and reuse of all wastes is economically sound, not only for society, but also for the involved industries themselves. All reuse begins with practices of waste minimization, whether by process change, direct selling to others, or indirect sale to regional exchanges.

The ultimate in all recovery and reuse operations then, is to reorganize our industrial production system to include balanced environmental complexes. As discussed earlier, these complexes will include two or more industrial plants, located adjacent to one another, so that all the wastes of one serve as part of the raw material for another. In this way, no gaseous, liquid, or solid wastes leave the complex to adversely affect the environment, and the production costs of all plants are minimized by their not having to transport or store raw materials or treat wastes.

INDEX